Springer Undergraduate Mathematics Series

The Springer Undergraduate Mathematics Series (SUMS) is a series designed for undergraduates in mathematics and the sciences worldwide. From core foundational material to final year topics, SUMS books take a fresh and modern approach. Textual explanations are supported by a wealth of examples, problems and fully-worked solutions, with particular attention paid to universal areas of difficulty. These practical and concise texts are designed for a one- or two-semester course but the self-study approach makes them ideal for independent use.

Nicolas Privault

Discrete Stochastic Processes

Tools for Machine Learning and Data Science

 Springer

Nicolas Privault
Division of Mathematical Sciences
Nanyang Technological University
Singapore, Singapore

ISSN 1615-2085   ISSN 2197-4144   (electronic)
Springer Undergraduate Mathematics Series
ISBN 978-3-031-65819-8   ISBN 978-3-031-65820-4   (eBook)
https://doi.org/10.1007/978-3-031-65820-4

This Springer imprint is published by the registered company Springer Nature Switzerland AG
The registered company address is: Gewerbestrasse 11, 6330 Cham, Switzerland

If disposing of this product, please recycle the paper.

Preface

This book aims at providing foundations in random processes for the understanding of machine learning and data science algorithms that revolve around the discrete-time Markov property. This includes mastering basic concepts in stochastic modeling for the understanding of topics such as synchronizing automata, the Markov Chain Monte Carlo (MCMC) method, statistical mechanics models, search engines, hidden Markov models, and reinforcement learning by Markov decision processes (MDP), for data science (Chaps. 2, 9, and 10), computer science/machine learning (Chaps. 3, 6, and 8), and applied sciences/physics (Chaps. 4, 5, and 7).

Those topics are considered from the angle of discrete-time stochastic processes, which are a central tool in this exposition, with selected applications involving random interactions that revolve around the Markov property. In particular, we cover excited random walks and related recurrence questions, distribution modeling using phase-type distributions, convergence and mixing of Markov chains, applications to search engines and probabilistic automata, and an introduction to the Ising model used in statistical physics. Applications to data science are treated via hidden Markov models and Markov decision processes.

The target audience of this book is the advanced undergraduate student in a quantitative field, mainly assuming that the reader has taken a first course in linear algebra, and a first course in probability and statistics, covering basic concepts such as conditional expectations and conditional probabilities. Prior knowledge of stochastic processes is not required, as the necessary prerequisites on Markov chains are recalled in Chap. 1.

The review presented in Chap. 1 is followed by applications to phase-type distributions and synchronizing automata in Chaps. 2 and 3, respectively, that can be used as illustrative examples for a better understanding of the Markov property. Random walks and their recurrence properties are considered in Chap. 4, with an extension to cookie-excited random walks that consider possible interaction with their environment in Chap. 5.

In Chap. 6 we consider the long-run behavior of Markov chains and present the Markov Chain Monte Carlo method which has multiple applications in biology, chemistry, physics, and computer science.

Next, in Chap. 7 we study the Ising model due to its applications in statistical mechanics and to complex random networks such as the ones generated by social media. The design of search engines considered in Chap. 8 also makes use of the results on convergence to equilibrium presented in Chap. 6.

The hidden Markov models treated in Chap. 9 have applications to, e.g., natural language processing (NLP), and the Markov decision processes (MDP) of Chap. 10 are used in reinforcement learning.

The following diagram shows the dependencies between the different chapters of the book.

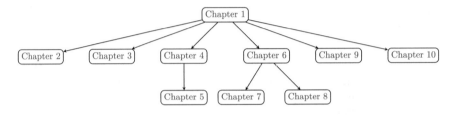

Application examples are presented via experiments and simulations based on 19 ℝ codes, 10 Python codes, and 5 animated figures available at https://github.com/nprivaul/discrete_stochastic_modeling, in addition to 48 figures and 3 tables.

The material in this book has been used for the Master of Science in Analytics at the Nanyang Technological University in Singapore, and for a GIAN course at the Indian Institute of Technology Madras at the invitation of Dr Neelesh Upadhye. This text also contains 31 original exercises and 17 new longer problems whose solutions are completely worked out in a supplementary solutions manual.

Singapore, Singapore   Nicolas Privault
June 2024

Contents

List of Figures

Chapter 1
A Summary of Markov Chains

This chapter reviews the concepts of discrete-time Markov process and matrix-based transition probabilities, which are central tools in this book. We also cover related techniques for the computation of hitting probabilities and mean hitting and absorption times, which will be applied in subsequent chapters. This chapter is mostly self-contained, to the exception of some proofs for which the reader is referred for conciseness to the relevant statements in the literature.

1.1 Markov Property

We let $\mathbb{N} = \{0, 1, 2, \ldots\}$ denote the set of non-negative integers, and consider a discrete-time stochastic process $(Z_n)_{n \geqslant 0}$ taking values in a countable discrete state space $\mathbb{S}$, typically $\mathbb{S} = \mathbb{Z}$. The $\mathbb{S}$-valued process $(Z_n)_{n \geqslant 0}$ is said to be *Markov*, or to have the *Markov property*, if, for all $n \geqslant 1$, the probability distribution of Z_{n+1} is determined by the state Z_n of the process at time n, and does not depend on the past values of Z_k for $k = 0, 1, \ldots, n - 1$.

In other words, for all $n \geqslant 1$ and all $i_0, i_1, \ldots, i_n, j \in \mathbb{S}$ we have

$$\mathbb{P}(Z_{n+1} = j \mid Z_n = i_n, \ Z_{n-1} = i_{n-1}, \ldots, \ Z_0 = i_0) = \mathbb{P}(Z_{n+1} = j \mid Z_n = i_n).$$

In particular, we have

$$\mathbb{P}(Z_{n+1} = j \mid Z_n = i_n, \ Z_{n-1} = i_{n-1}) = \mathbb{P}(Z_{n+1} = j \mid Z_n = i_n),$$

and, for $n = 1$,

$$\mathbb{P}(Z_2 = j \mid Z_1 = i_1, Z_0 = i_0) = \mathbb{P}(Z_2 = j \mid Z_1 = i_1).$$

In addition, we have the following facts.

© The Author(s), under exclusive license to Springer Nature Switzerland AG 2024
N. Privault, *Discrete Stochastic Processes*, Springer Undergraduate
Mathematics Series, https://doi.org/10.1007/978-3-031-65820-4_1

1. *Chain rule.* The first order transition probabilities can be used for the complete computation of the probability distribution of the process $(Z_n)_{n \geqslant 0}$ by induction, as

$$\mathbb{P}(Z_n = i_n, Z_{n-1} = i_{n-1}, \dots, Z_0 = i_0) \tag{1.1}$$
$$= \mathbb{P}(Z_n = i_n \mid Z_{n-1} = i_{n-1}) \cdots \mathbb{P}(Z_1 = i_1 \mid Z_0 = i_0)\mathbb{P}(Z_0 = i_0),$$

or, after dividing both sides by $\mathbb{P}(Z_0 = i_0)$,

$$\mathbb{P}(Z_n = i_n, Z_{n-1} = i_{n-1}, \dots, Z_1 = i_1 \mid Z_0 = i_0) \tag{1.2}$$
$$= \mathbb{P}(Z_n = i_n \mid Z_{n-1} = i_{n-1}) \cdots \mathbb{P}(Z_1 = i_1 \mid Z_0 = i_0),$$

$i_0, i_1, \dots, i_n \in \mathbb{S}$.

2. By the *law of total probability* applied under $\mathbb{P}$ to the events

$$A_{i_0} := \{Z_1 = i_1 \text{ and } Z_0 = i_0\}, \qquad i_0 \in \mathbb{S},$$

we have

$$\mathbb{P}(Z_1 = i_1) = \mathbb{P}\left(\bigcup_{i_0 \in \mathbb{S}} \{Z_1 = i_1, Z_0 = i_0\}\right)$$
$$= \sum_{i_0 \in \mathbb{S}} \mathbb{P}(Z_1 = i_1, Z_0 = i_0)$$
$$= \sum_{i_0 \in \mathbb{S}} \mathbb{P}(Z_1 = i_1 \mid Z_0 = i_0)\mathbb{P}(Z_0 = i_0), \quad i_1 \in \mathbb{S}. \tag{1.3}$$

Similarly, under the probability measure $\mathbb{P}(\cdot \mid Z_0 = i_0)$, we have

$$\mathbb{P}(Z_2 = i_2 \mid Z_0 = i_0) = \mathbb{P}\left(\bigcup_{i_1 \in \mathbb{S}} \{Z_2 = i_2 \text{ and } Z_1 = i_1\} \,\middle|\, Z_0 = i_0\right)$$
$$= \sum_{i_1 \in \mathbb{S}} \mathbb{P}(Z_2 = i_2 \text{ and } Z_1 = i_1 \mid Z_0 = i_0)$$
$$= \sum_{i_1 \in \mathbb{S}} \mathbb{P}(Z_2 = i_2 \mid Z_1 = i_1)\mathbb{P}(Z_1 = i_1 \mid Z_0 = i_0), \quad i_0, i_2 \in \mathbb{S}. \tag{1.4}$$

Transition Matrices

In what follows, we will make the following assumption.

Assumption (A) *The Markov chain* $(Z_n)_{n \geqslant 0}$ *is* time homogeneous, *i.e. the transition probabilities*

$$\mathbb{P}(Z_{n+1} = j \mid Z_n = i), \qquad i, j \in S,$$

do not depend on $n \geqslant 0$.

Under Assumption (A) the random evolution of a Markov chain $(Z_n)_{n \geqslant 0}$ is determined by the data of

$$P_{i,j} := \mathbb{P}(Z_1 = j \mid Z_0 = i), \qquad i, j \in S, \tag{1.5}$$

which coincides with the probability $\mathbb{P}(Z_{n+1} = j \mid Z_n = i)$ for all $n \in \mathbb{N}$. These data can be encoded into a matrix indexed by $S^2 = S \times S$, called the *transition matrix* of the Markov chain:

$$\left[P_{i,j} \right]_{i,j \in S} = \left[\mathbb{P}(Z_1 = j \mid Z_0 = i) \right]_{i,j \in S},$$

also written on $S := \mathbb{Z}$ as

$$P = \left[P_{i,j} \right]_{i,j \in S} = \begin{bmatrix} \ddots & \vdots & \vdots & \vdots & \vdots & \vdots & \iddots \\ \cdots & P_{-2,-2} & P_{-2,-1} & P_{-2,0} & P_{-2,1} & P_{-2,2} & \cdots \\ \cdots & P_{-1,-2} & P_{-1,-1} & P_{-1,0} & P_{-1,1} & P_{-1,2} & \cdots \\ \cdots & P_{0,-2} & P_{0,-1} & P_{0,0} & P_{0,1} & P_{0,2} & \cdots \\ \cdots & P_{1,-2} & P_{1,-1} & P_{1,0} & P_{1,1} & P_{1,2} & \cdots \\ \cdots & P_{2,-2} & P_{2,-1} & P_{2,0} & P_{2,1} & P_{2,2} & \cdots \\ \iddots & \vdots & \vdots & \vdots & \vdots & \vdots & \ddots \end{bmatrix}.$$

By the *law of total probability* applied to the probability measure $\mathbb{P}(\cdot \mid Z_0 = i)$, we also have the equality

$$\sum_{j \in S} \mathbb{P}(Z_1 = j \mid Z_0 = i) = \mathbb{P}\left(\bigcup_{j \in S} \{Z_1 = j\} \,\middle|\, Z_0 = i \right) = \mathbb{P}(\Omega) = 1, \quad i \in S, \tag{1.6}$$

i.e. the *rows* of the transition matrix satisfy the condition

$$\sum_{j \in S} P_{i,j} = 1,$$

for every row index $i \in S$.

Using the matrix notation $P = (P_{i,j})_{i,j \in S}$ and Relation (1.1), we find

$$\mathbb{P}(Z_n = i_n, Z_{n-1} = i_{n-1}, \ldots, Z_0 = i_0) = P_{i_{n-1}, i_n} \cdots P_{i_0, i_1} \mathbb{P}(Z_0 = i_0),$$

$i_0, i_1, \ldots, i_n \in S$, and we rewrite (1.3) as

$$\mathbb{P}(Z_1 = i) = \sum_{j \in S} \mathbb{P}(Z_1 = i \mid Z_0 = j)\mathbb{P}(Z_0 = j) = \sum_{j \in S} P_{j,i} \mathbb{P}(Z_0 = j), \quad i \in S.$$

$$(1.7)$$

A state $k \in S$ is said to be *absorbing* if $P_{k,k} = 1$.

In case the Markov chain $(Z_k)_{k \in \mathbb{N}}$ takes values in the finite state space $S = \{0, 1, \ldots, N\}$, its $(N + 1) \times (N + 1)$ transition matrix will simply have the form

$$\left[P_{i,j} \right]_{0 \leqslant i,j \leqslant N} = \begin{bmatrix} P_{0,0} & P_{0,1} & P_{0,2} & \cdots & P_{0,N} \\ P_{1,0} & P_{1,1} & P_{1,2} & \cdots & P_{1,N} \\ P_{2,0} & P_{2,1} & P_{2,2} & \cdots & P_{2,N} \\ \vdots & \vdots & \vdots & \ddots & \vdots \\ P_{N,0} & P_{N,1} & P_{N,2} & \cdots & P_{N,N} \end{bmatrix}.$$

Still on the finite state space $S = \{0, 1, \ldots, N\}$, Relations (1.3) and (1.7) can be restated in the language of matrix and vector products using the shorthand notation

$$\eta = \pi P, \tag{1.8}$$

where

$$\eta := [\mathbb{P}(Z_1 = 0), \ldots, \mathbb{P}(Z_1 = N)] = [\eta_0, \eta_1, \ldots, \eta_N] \in \mathbb{R}^{N+1}$$

is the row vector "distribution of Z_1",

$$\pi := [\mathbb{P}(Z_0 = 0), \ldots, \mathbb{P}(Z_0 = N)] = [\pi_0, \ldots, \pi_N] \in \mathbb{R}^{N+1}$$

is the row vector representing the probability distribution of Z_0, and

$$[\eta_0, \eta_1, \ldots, \eta_N] = [\pi_0, \ldots, \pi_N] \times \begin{bmatrix} P_{0,0} & P_{0,1} & P_{0,2} & \cdots & P_{0,N} \\ P_{1,0} & P_{1,1} & P_{1,2} & \cdots & P_{1,N} \\ P_{2,0} & P_{2,1} & P_{2,2} & \cdots & P_{2,N} \\ \vdots & \vdots & \vdots & \ddots & \vdots \\ P_{N,0} & P_{N,1} & P_{N,2} & \cdots & P_{N,N} \end{bmatrix}. \tag{1.9}$$

The ® code_01_page_5.R illustrates the use of transition matrices for the modeling of Markov chains

```
1   install.packages("devtools"); library(devtools) # Install RTools as well
    devtools::install_github('spedygiorgio/markovchain')
3   install.packages("igraph"); library(igraph); library(markovchain)
    P<-matrix(c(1,0,0,0,1./3,0,1./3,1./3,1./3,1./3,0,1./3,0,0,0,1),nrow=4, byrow=TRUE)
5   MC <-new("markovchain", transitionMatrix=P, states=c("0","1","2","3"))
    graph <- as(MC, "igraph")
7   plot(graph,edge.label.cex=0.8,edge.label=sprintf("%1.2f", E(graph)$prob),
        edge.color='black', vertex.color='dodgerblue', vertex.label.cex=0.8)
```

Higher-Order Transition Probabilities

As noted above, the transition matrix $P = (P_{i,j})_{i,j \in S}$ represents a convenient way to record $\mathbb{P}(Z_{n+1} = j \mid Z_n = i)$, $i, j \in S$, into an array of data.

However, it is *much more than that*, as already hinted at in Relation (1.8). Suppose for example that we are interested in the two-step transition probability

$$\mathbb{P}(Z_{n+2} = j \mid Z_n = i).$$

This probability does not appear in the transition matrix P, but it can be computed by first step analysis, applying the *law of total probability* to the conditional probability measure $\mathbb{P}(\cdot \mid Z_n = i)$.

(i) 2-step transitions. Denoting by S the state space of the process we have, using (1.4),

$$\mathbb{P}(Z_{n+2} = j \mid Z_n = i) = \sum_{l \in S} \mathbb{P}(Z_{n+2} = j \text{ and } Z_{n+1} = l \mid Z_n = i)$$

$$= \sum_{l \in S} \mathbb{P}(Z_{n+2} = j \mid Z_{n+1} = l)\mathbb{P}(Z_{n+1} = l \mid Z_n = i)$$

$$= \sum_{l \in S} P_{i,l} P_{l,j}$$

$$= [P^2]_{i,j}, \qquad i, j \in S,$$

where we used (1.5), which is in agreement with the matrix multiplication mechanism described below.

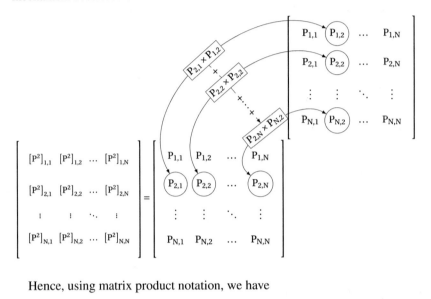

Hence, using matrix product notation, we have

$$(\mathbb{P}(Z_{n+2} = j \mid Z_n = i))_{0 \leqslant i, j \leqslant N}$$

$$= \begin{bmatrix} P_{0,0} & P_{0,1} & P_{0,2} & \cdots & P_{0,N} \\ P_{1,0} & P_{1,1} & P_{1,2} & \cdots & P_{1,N} \\ P_{2,0} & P_{2,1} & P_{2,2} & \cdots & P_{2,N} \\ \vdots & \vdots & \vdots & \ddots & \vdots \\ P_{N,0} & P_{N,1} & P_{N,2} & \cdots & P_{N,N} \end{bmatrix} \times \begin{bmatrix} P_{0,0} & P_{0,1} & P_{0,2} & \cdots & P_{0,N} \\ P_{1,0} & P_{1,1} & P_{1,2} & \cdots & P_{1,N} \\ P_{2,0} & P_{2,1} & P_{2,2} & \cdots & P_{2,N} \\ \vdots & \vdots & \vdots & \ddots & \vdots \\ P_{N,0} & P_{N,1} & P_{N,2} & \cdots & P_{N,N} \end{bmatrix}.$$

(ii) k-step transitions. More generally, we have the following result.

Proposition 1.1 *For all $k \in \mathbb{N}$ we have the relation*

$$\left[\mathbb{P}(Z_{n+k} = j \mid Z_n = i) \right]_{i,j \in S} = \left[[P^k]_{i,j} \right]_{i,j \in S} = P^k. \tag{1.10}$$

Proof We prove (1.10) by induction. Clearly, the statement holds for $k = 0$ and $k = 1$. Next, for all $k \in \mathbb{N}$, we have

$$\mathbb{P}(Z_{n+k+1} = j \mid Z_n = i) = \sum_{l \in S} \mathbb{P}(Z_{n+k+1} = j \text{ and } Z_{n+k} = l \mid Z_n = i)$$

$$= \sum_{l \in S} \frac{\mathbb{P}(Z_{n+k+1} = j, \ Z_{n+k} = l, \ Z_n = i)}{\mathbb{P}(Z_n = i)}$$

$$= \sum_{l \in S} \frac{\mathbb{P}(Z_{n+k+1} = j, \ Z_{n+k} = l, \ Z_n = i)}{\mathbb{P}(Z_{n+k} = l \text{ and } Z_n = i)} \frac{\mathbb{P}(Z_{n+k} = l \text{ and } Z_n = i)}{\mathbb{P}(Z_n = i)}$$

$$= \sum_{l \in S} \mathbb{P}(Z_{n+k+1} = j \mid Z_{n+k} = l \text{ and } Z_n = i) \mathbb{P}(Z_{n+k} = l \mid Z_n = i)$$

$$= \sum_{l \in S} \mathbb{P}(Z_{n+k+1} = j \mid Z_{n+k} = l) \mathbb{P}(Z_{n+k} = l \mid Z_n = i)$$

$$= \sum_{l \in S} \mathbb{P}(Z_{n+k} = l \mid Z_n = i) P_{l,j}.$$

We have just checked that the family of matrices

$$[\, \mathbb{P}(Z_{n+k} = j \mid Z_n = i) \,]_{i,j \in S}, \qquad k \geqslant 1,$$

satisfies the same induction relation as the *matrix power P^k*, i.e.

$$[P^{k+1}]_{i,j} = \sum_{l \in S} [P^k]_{i,l} P_{l,j},$$

and the same initial condition, hence by induction on $k \geqslant 0$ the equality

$$[\, \mathbb{P}(Z_{n+k} = j \mid Z_n = i) \,]_{i,j \in S} = \left[[P^k]_{i,j} \right]_{i,j \in S} = P^k$$

holds not only for $k = 0$ and $k = 1$, but also for all $k \in \mathbb{N}$. $\qquad\square$

The matrix product relation

$$P^{m+n} = P^m P^n = P^n P^m,$$

reads

$$[P^{m+n}]_{i,j} = \sum_{l \in S} [P^m]_{i,l} [P^n]_{l,j} = \sum_{l \in S} [P^n]_{i,l} [P^m]_{l,j}, \qquad i, j \in S,$$

and can now be interpreted as

$$\mathbb{P}(Z_{n+m} = j \mid Z_0 = i) = \sum_{l \in S} \mathbb{P}(Z_{n+m} = j \mid Z_n = l) \mathbb{P}(Z_n = l \mid Z_0 = i)$$

$$= \sum_{l \in S} \mathbb{P}(Z_m = j \mid Z_0 = l) \mathbb{P}(Z_n = l \mid Z_0 = i)$$

$$= \sum_{l \in S} \mathbb{P}(Z_n = j \mid Z_0 = l) \mathbb{P}(Z_m = l \mid Z_0 = i), \qquad i, j \in S,$$

which is called the *Chapman-Kolmogorov* equation.

1.2   Hitting Probabilities

Starting with this section, we introduce the systematic use of the first step analysis technique. The main applications of first step analysis are the computation of hitting probabilities, mean hitting and absorption times, mean first return times, and average number of returns to a given state.

Hitting Probabilities

Let us consider a Markov chain $(Z_n)_{n \geq 0}$ with state space S, and let $\mathcal{A} \subset S$ denote a subset of S as in the following example with $S = \{0, 1, 2, 3, 4, 5\}$ and $\mathcal{A} := \{0, 2, 4\}$, with

$$P_{k,l} = \mathbb{1}_{\{k=l\}} \quad \text{for all} \quad k, l \in \mathcal{A}, \tag{1.11}$$

in which case the set $\mathcal{A} \subset S$ is said to be *absorbing*.

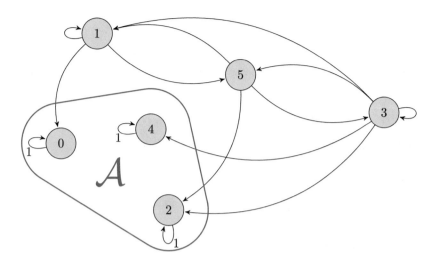

We are interested in the first time T_A the chain hits the subset A, with

$$T_A := \inf\{n \geqslant 0 \; : \; Z_n \in A\}, \tag{1.12}$$

with $T_A = 0$ if $Z_0 \in A$ and

$$T_A = \infty \quad \text{if} \quad \{n \geqslant 0 \; : \; Z_n \in A\} = \emptyset,$$

i.e. if $Z_n \notin A$ for all $n \in \mathbb{N}$.

We now aim at computing the hitting probabilities

$$g_l(k) = \mathbb{P}(Z_{T_A} = l \text{ and } T_A < \infty \mid Z_0 = k)$$

of hitting the set $A \subset S$ through state $l \in A$ starting from $k \in S$, where Z_{T_A} represents the location of the chain $(Z_n)_{n \geqslant 0}$ at the hitting time T_A. This computation can be achieved by first step analysis, using the *law of total probability* for the probability measure $\mathbb{P}(\cdot \mid Z_0 = k)$ and the Markov property, as follows.

Proposition 1.2 *Assume that* (1.11) *holds. The hitting probabilities*

$$g_l(k) := \mathbb{P}(Z_{T_A} = l \text{ and } T_A < \infty \mid Z_0 = k), \qquad k \in S, \; l \in A,$$

satisfy the equation

$$g_l(k) = \sum_{m \in S} P_{k,m} g_l(m) = P_{k,l} + \sum_{m \in S \setminus A} P_{k,m} g_l(m), \tag{1.13}$$

k ∈ S \ A, *l* ∈ A, *under the boundary conditions*

$$g_l(k) = \mathbb{P}(Z_{T_A} = l \text{ and } T_A < \infty \mid Z_0 = k) = \mathbb{1}_{\{k=l\}} = \begin{cases} 1 \text{ if } k = l, \\ \\ 0 \text{ if } k \neq l, \end{cases} \quad k, l \in A,$$

which hold since $T_A = 0$ *whenever one starts from* $Z_0 \in A$.

Proof For all $k \in S \setminus A$ we have $T_A \geqslant 1$ given that $Z_0 = k$, hence we can write

$$g_l(k) = \mathbb{P}(Z_{T_A} = l \text{ and } T_A < \infty \mid Z_0 = k)$$

$$= \sum_{m \in S} \mathbb{P}(Z_{T_A} = l \text{ and } T_A < \infty \mid Z_1 = m \text{ and } Z_0 = k)\mathbb{P}(Z_1 = m \mid Z_0 = k)$$

$$= \sum_{m \in S} \mathbb{P}(Z_{T_A} = l \text{ and } T_A < \infty \mid Z_1 = m)\mathbb{P}(Z_1 = m \mid Z_0 = k)$$

$$= \sum_{m \in S} P_{k,m}\mathbb{P}(Z_{T_A} = l \text{ and } T_A < \infty \mid Z_1 = m)$$

$$= \sum_{m \in S} P_{k,m}\mathbb{P}(Z_{T_A} = l \text{ and } T_A < \infty \mid Z_0 = m)$$

$$= \sum_{m \in S} P_{k,m} g_l(m), \qquad k \in S \setminus A, \quad l \in A,$$

where the relation

$$\mathbb{P}(Z_{T_A} = l \text{ and } T_A < \infty \mid Z_1 = m) = \mathbb{P}(Z_{T_A} = l \text{ and } T_A < \infty \mid Z_0 = m)$$

follows from the fact that this hitting probability does not depend on the initial time the counter is started, as the chain is *time homogeneous*.   □

Remarks

- See e.g. Theorem 3.4 page 40 of Karlin and Taylor (1981) for a uniqueness result for the solution of such equations, and Theorem 2.1 in Goldberg (1986) for the uniqueness of solutions to difference equations in general.
- The commands `absorbingStates(MC)` and `hittingProbabilities(MC)` in the ℝ markovchain package can be used together with the ℝ code_01_page_5.R to determine the absorbing states and their hitting probabilities.
- Equation (1.13) can be rewritten in matrix form as

$$g_l = P g_l, \qquad l \in A,$$

where g_l is a column vector, i.e.

$$
\begin{bmatrix} g_l(0) \\ \vdots \\ g_l(N) \end{bmatrix} = \begin{bmatrix} P_{0,0} & P_{0,1} & P_{0,2} & \cdots & P_{0,N} \\ P_{1,0} & P_{1,1} & P_{1,2} & \cdots & P_{1,N} \\ P_{2,0} & P_{2,1} & P_{2,2} & \cdots & P_{2,N} \\ \vdots & \vdots & \vdots & \ddots & \vdots \\ P_{N,0} & P_{N,1} & P_{N,2} & \cdots & P_{N,N} \end{bmatrix} \times \begin{bmatrix} g_l(0) \\ \vdots \\ g_l(N) \end{bmatrix}, \qquad l \in \mathcal{A},
$$

under the boundary condition

$$
g_l(k) = \mathbb{P}(Z_{T_\mathcal{A}} = l \text{ and } T_\mathcal{A} < \infty \mid Z_0 = k) = \mathbb{1}_{\{l\}}(k) = \begin{cases} 1, & k = l, \\ 0, & k \neq l, \end{cases}
$$

for all $k, l \in \mathcal{A}$.

- The hitting probabilities $g_l(k) = \mathbb{P}(Z_{T_\mathcal{A}} = l \text{ and } T_\mathcal{A} < \infty \mid Z_0 = k)$ satisfy the condition

$$
1 = \mathbb{P}(T_\mathcal{A} = \infty \mid Z_0 = k) + \sum_{l \in \mathcal{A}} \mathbb{P}(Z_{T_\mathcal{A}} = l \text{ and } T_\mathcal{A} < \infty \mid Z_0 = k)
$$

$$
= \mathbb{P}(T_\mathcal{A} = \infty \mid Z_0 = k) + \sum_{l \in \mathcal{A}} g_l(k), \tag{1.14}
$$

for all $k \in \mathbb{S}$.

- Note that we may have $\mathbb{P}(T_\mathcal{A} = \infty \mid Z_0 = k) > 0$, for example in the following chain with $\mathcal{A} = \{0\}$ and $k = 1$ we have

$$
\mathbb{P}(T_0 = \infty \mid Z_0 = 1) = 0.2.
$$

- Consider $f : \mathcal{A} \longrightarrow \mathbb{R}$ a function on the domain $\mathcal{A}$, and assume that $\mathbb{P}(T_\mathcal{A} < \infty \mid Z_0 = k) = 1$, $k \in \mathbb{S}$. Letting

$$
g_\mathcal{A}(k) := \mathbb{E}[f(Z_{T_\mathcal{A}}) \mid Z_0 = k] = \sum_{l \in \mathcal{A}} f(l) \mathbb{P}(Z_{T_\mathcal{A}} = l \mid Z_0 = k), \quad k \in \mathbb{S},
$$

the first step analysis argument of Proposition 1.2 can be used to show that g_A solves the *Dirichlet problem*

$$g_A = Pg_A,$$

with boundary condition

$$g_A(k) = f(k), \qquad k \in A.$$

1.3 Mean Hitting and Absorption Times

We are now interested in the mean hitting time $h_A(k)$ it takes for the chain to hit the set $A \subset S$ starting from a state $k \in S$. This mean hitting time is defined as the conditional expectation

$$h_A(k) := \mathbb{E}[T_A \mid Z_0 = k] = \frac{1}{\mathbb{P}(Z_0 = k)} \mathbb{E}[T_A \mathbb{1}_{\{Z_0 = k\}}], \quad k \in S. \tag{1.15}$$

In case the set A is absorbing, we refer to $h_A(k)$ as the *mean absorption time* into A starting from the state $\widehat{k}$. Clearly, since $T_A = 0$ whenever $Z_0 = k \in A$, we have

$$h_A(k) = 0, \qquad \text{for all } k \in A.$$

Proposition 1.3 *The mean hitting times* (1.15) *satisfy the equations*

$$h_A(k) = 1 + \sum_{l \in S} P_{k,l} h_A(l) = 1 + \sum_{l \in S \setminus A} P_{k,l} h_A(l), \quad k \in S \setminus A, \tag{1.16}$$

under the boundary conditions

$$h_A(k) = \mathbb{E}[T_A \mid Z_0 = k] = 0, \qquad k \in A.$$

Proof For all $k \in S \setminus A$, by first step analysis using the *law of total expectation* applied to the probability measure $\mathbb{P}(\cdot \mid Z_0 = l)$, and the Markov property we have

$$h_A(k) = \mathbb{E}[T_A \mid Z_0 = k]$$

$$= \sum_{l \in S} \mathbb{E}[1 + T_A \mid Z_0 = l] \mathbb{P}(Z_1 = l \mid Z_0 = k)$$

$$= \sum_{l \in S} (1 + \mathbb{E}[T_A \mid Z_0 = l]) \mathbb{P}(Z_1 = l \mid Z_0 = k)$$

$$= \sum_{l \in S} \mathbb{P}(Z_1 = l \mid Z_0 = k) + \sum_{l \in S} \mathbb{P}(Z_1 = l \mid Z_0 = k)\mathbb{E}[T_{\mathcal{A}} \mid Z_0 = l]$$

$$= 1 + \sum_{l \in S} \mathbb{P}(Z_1 = l \mid Z_0 = k)\mathbb{E}[T_{\mathcal{A}} \mid Z_0 = l]$$

$$= 1 + \sum_{l \in S} P_{k,l} h_{\mathcal{A}}(l), \qquad k \in S \setminus \mathcal{A}.$$

Hence, we have

$$h_{\mathcal{A}}(k) = 1 + \sum_{l \in S} P_{k,l} h_{\mathcal{A}}(l), \quad k \in S \setminus \mathcal{A}, \tag{1.17}$$

under the boundary conditions

$$h_{\mathcal{A}}(k) = \mathbb{E}[T_{\mathcal{A}} \mid Z_0 = k] = 0, \qquad k \in \mathcal{A}, \tag{1.18}$$

the Condition (1.18) implies that (1.17) becomes

$$h_{\mathcal{A}}(k) = 1 + \sum_{l \in S \setminus \mathcal{A}} P_{k,l} h_{\mathcal{A}}(l), \qquad k \in S \setminus \mathcal{A}.$$

□

The command `meanAbsorptionTime(MC)` can be used together with the ℝ code_01_page_5.R used to determine mean absorption times. The Eq. (1.16) can be rewritten in matrix form as

$$h_{\mathcal{A}} = \begin{bmatrix} 1 \\ \vdots \\ 1 \end{bmatrix} + P h_{\mathcal{A}},$$

by considering only the rows with index $k \in \mathcal{A}^c = S \setminus \mathcal{A}$, under the boundary conditions

$$h_{\mathcal{A}}(k) = 0, \qquad k \in \mathcal{A}.$$

First Return Times

Consider now the first *return* time T_j^r to state $j \in S$, defined by

$$T_j^r := \inf\{n \geqslant 1 \; : \; Z_n = j\},$$

with

$$T_j^r = \infty \quad \text{if } Z_n \neq j \text{ for all } n \geqslant 1.$$

Note that in contrast with the definition (1.12) of the hitting time T_j, the infimum is taken here for $n \geqslant 1$ as it takes at least one step out of the initial state in order to *return* to state $\textcircled{j}$. Nevertheless we have $T_j = T_j^r$ when the chain is started from a state $\textcircled{i}$ different from $\textcircled{j}$. We denote by

$$\mu_j(i) := \mathbb{E}\big[T_j^r \,\big|\, Z_0 = i\big] \geqslant 1$$

the *mean return time* to state $j \in S$ after starting from state $i \in S$. Mean return times can also be computed by first step analysis, as follows. We have

$$
\begin{aligned}
\mu_j(i) &= \mathbb{E}\big[T_j^r \mid Z_0 = i\big] \\
&= 1 \times \mathbb{P}(Z_1 = j \mid Z_0 = i) \\
&\quad + \sum_{\substack{l \in S \\ l \neq j}} \mathbb{P}(Z_1 = l \mid Z_0 = i)\big(1 + \mathbb{E}\big[T_j^r \mid Z_0 = l\big]\big) \\
&= P_{i,j} + \sum_{\substack{l \in S \\ l \neq j}} P_{i,l}(1 + \mu_j(l)) \\
&= P_{i,j} + \sum_{\substack{l \in S \\ l \neq j}} P_{i,l} + \sum_{\substack{l \in S \\ l \neq j}} P_{i,l}\mu_j(l) \\
&= \sum_{l \in S} P_{i,l} + \sum_{\substack{l \in S \\ l \neq j}} P_{i,l}\mu_j(l) \\
&= 1 + \sum_{\substack{l \in S \\ l \neq j}} P_{i,l}\mu_j(l),
\end{aligned}
$$

hence

$$\mu_j(i) = 1 + \sum_{\substack{l \in S \\ l \neq j}} P_{i,l}\mu_j(l), \qquad i, j \in S. \tag{1.19}$$

See e.g. Theorem 5.9 page 49 of Karlin and Taylor (1981) for a uniqueness result for the solution of such equations.

Hitting Times vs. Return Times

Note that the time T_i^r to return to state (i) is always at least one by construction, hence $\mu_i(i) \geqslant 1$ and cannot vanish, while we always have $h_i(i) = 0$ as boundary condition, $i \in S$. On the other hand, for $i \neq j$ we have by definition

$$h_i(j) = \mathbb{E}\big[T_i^r \mid Z_0 = j\big] = \mathbb{E}[T_i \mid Z_0 = j] = \mu_i(j),$$

and for $i = j$ the mean return time $\mu_j(j)$ can be computed from the hitting times $h_j(l), l \neq j$, by first step analysis as

$$\mu_j(j) = \sum_{l \in S} P_{j,l}(1 + h_j(l))$$

$$= P_{j,j} + \sum_{l \neq j} P_{j,l}(1 + h_j(l))$$

$$= \sum_{l \in S} P_{j,l} + \sum_{l \neq j} P_{j,l} h_j(l)$$

$$= 1 + \sum_{l \neq j} P_{j,l} h_j(l), \qquad j \in S, \tag{1.20}$$

which is in agreement with (1.19) when $i = j$.

Markov Chains with Rewards

Let $(Z_n)_{n \geqslant 0}$ be a Markov chain with state space S and transition matrix $P = (P_{i,j})_{i,j \in S}$. Derive the first step analysis equation for the value function

$$V(k) := \mathbb{E}\left[\sum_{n \geqslant 0} q^n R(Z_n) \,\bigg|\, Z_0 = k\right], \qquad k \in S, \tag{1.21}$$

defined as the total accumulated reward obtained after starting from state (k), where $R : S \to \mathbb{R}$ is a reward function and $q \in (0, 1]$ is a *discount factor*. We have

$$V(k) = \mathbb{E}\left[\sum_{n \geqslant 0} q^n R(Z_n) \,\bigg|\, Z_0 = k\right]$$

$$= \mathbb{E}[R(Z_0) \mid Z_0 = k] + \mathbb{E}\left[\sum_{n \geqslant 1} q^n R(Z_n) \,\bigg|\, Z_0 = k\right]$$

$$= R(k) + \sum_{m \in S} P_{k,m} \mathbb{E} \left[\sum_{n \geqslant 1} q^n R(Z_n) \;\middle|\; Z_1 = m \right]$$

$$= R(k) + q \sum_{m \in S} P_{k,m} \mathbb{E} \left[\sum_{n \geqslant 0} q^n R(Z_n) \;\middle|\; Z_0 = m \right]$$

$$= R(k) + q \sum_{m \in S} P_{k,m} V(m), \qquad k \in S. \tag{1.22}$$

On a state space $S = \{1, \ldots, d\}$, the above Eq. (1.22) rewrites in matrix form as

$$V = \begin{bmatrix} R(1) \\ \vdots \\ R(d) \end{bmatrix} + q P V.$$

The command `expectedRewards(MC,100,c(0,4,-3,0))` can be used together with the ®R code_02_page_16.R to compute expected rewards, where the sequence `c(0,4,-3,0)` represents the rewards assigned to states $1, 2, 3, 4 \in S$.

See Chap. 10 for exercises on the computation of expected rewards.

Mean Number of Returns

Let

$$R_j := \sum_{n \geqslant 1} \mathbb{1}_{\{Z_n = j\}} \tag{1.23}$$

denote the number of returns to state (j) by the chain $(Z_n)_{n \geqslant 0}$.

Definition 1.4 For $i, j \in S$, let

$$p_{ij} = \mathbb{P}\left(T_j^r < \infty \mid Z_0 = i\right) = \mathbb{P}(Z_n = j \text{ for some } n \geqslant 1 \mid Z_0 = i),$$

denote the probability of return to state (j) in finite time[1] starting from state (i).

Proposition 1.5 can be derived by a straightforward argument using the geometric distribution.

---

[1] When $(i) \neq (j)$, p_{ij} is the probability of *visiting* state (j) in finite time after starting from state (i).

Proposition 1.5 *The probability distribution of the number of returns R_j to state j given that $\{Z_0 = i\}$ is given by*

$$
\mathbb{P}(R_j = m \mid Z_0 = i) = \begin{cases} 1 - p_{ij}, & m = 0, \\[2ex] p_{ij} \times (p_{jj})^{m-1} \times (1 - p_{jj}), & m \geqslant 1, \end{cases}
$$

In case $i = j$, R_i is simply the number of returns to state $\textcircled{i}$ starting from state $\textcircled{i}$, and it has the geometric distribution

$$
\mathbb{P}(R_i = m \mid Z_0 = i) = (1 - p_{ii})(p_{ii})^m, \qquad m \geqslant 0. \tag{1.24}
$$

Proposition 1.6 *We have*

$$
\mathbb{P}(R_j < \infty \mid Z_0 = i) = \begin{cases} 1 - p_{ij}, & \text{if } p_{jj} = 1, \\[2ex] 1, & \text{if } p_{jj} < 1. \end{cases}
$$

Proof By Proposition 1.5, we have

$$
\mathbb{P}(R_j < \infty \mid Z_0 = i) = \sum_{m \geqslant 0} \mathbb{P}(R_j = m \mid Z_0 = i)
$$

$$
= 1 - p_{ij} + (1 - p_{jj})p_{ij} \sum_{m \geqslant 1} (p_{jj})^{m-1}
$$

$$
= \begin{cases} 1 - p_{ij}, & \text{if } p_{jj} = 1, \\[2ex] 1, & \text{if } p_{jj} < 1. \end{cases}
$$

$\square$

Remarks

- As a consequence of Proposition 1.6, we also have

$$
\mathbb{P}(R_j = \infty \mid Z_0 = i) = \begin{cases} p_{ij}, & \text{if } p_{jj} = 1, \\[2ex] 0, & \text{if } p_{jj} < 1. \end{cases}
$$

- In particular, if $p_{jj} = 1$, i.e. state $\textcircled{j}$ is recurrent, we have

$$
\mathbb{P}(R_j = m \mid Z_0 = i) = 0, \qquad m \geqslant 1,
$$

and in this case, by Proposition 1.6 we have

$$
\begin{cases}
\mathbb{P}(R_j < \infty \mid Z_0 = i) = \mathbb{P}(R_j = 0 \mid Z_0 = i) = 1 - p_{ij}, \\[2mm]
\mathbb{P}(R_j = \infty \mid Z_0 = i) = 1 - \mathbb{P}(R_j < \infty \mid Z_0 = i) = p_{ij}.
\end{cases}
\tag{1.25}
$$

- On the other hand, when $i = j$ we find

$$
\begin{aligned}
\mathbb{P}(R_i < \infty \mid Z_0 = i) &= \sum_{m \geqslant 0} \mathbb{P}(R_i = m \mid Z_0 = i) \\
&= (1 - p_{ii}) \sum_{m \geqslant 0} (p_{ii})^m \\
&= \begin{cases}
0, & \text{if } p_{ii} = 1, \\
1, & \text{if } p_{ii} < 1,
\end{cases}
\end{aligned}
\tag{1.26}
$$

hence

$$
\mathbb{P}(R_i = \infty \mid Z_0 = i) = \begin{cases}
1, & \text{if } p_{ii} = 1, \\
0, & \text{if } p_{ii} < 1,
\end{cases}
\tag{1.27}
$$

i.e. the number of returns to a recurrent state is infinite with probability one.

The notion of *mean number of returns* will be needed for the classification of states of Markov chains in Sect. 1.4.

Proposition 1.7 *Assume that* $p_{ij} > 0$. *The mean number of returns to state* $\boxed{j}$ *is given by*

$$
\mathbb{E}[R_j \mid Z_0 = i] = \frac{p_{ij}}{1 - p_{jj}},
$$

and it is finite, i.e. $\mathbb{E}[R_j \mid Z_0 = i] < \infty$, *if and only if* $p_{jj} < 1$, $i, j \in S$.

Proof By (B.4), when $p_{jj} < 1$ we have $\mathbb{P}(R_j < \infty \mid Z_0 = i) = 1$ and

$$
\begin{aligned}
\mathbb{E}[R_j \mid Z_0 = i] &= \sum_{m \geqslant 1} m \mathbb{P}(R_j = m \mid Z_0 = i) \tag{1.28} \\
&= (1 - p_{jj}) p_{ij} \sum_{m \geqslant 1} m (p_{jj})^{m-1} \\
&= \frac{p_{ij}}{1 - p_{jj}}, \tag{1.29}
\end{aligned}
$$

see Relation (B.4), hence

$$\mathbb{E}[R_j \mid Z_0 = i] < \infty \quad \text{if} \quad p_{jj} < 1.$$

If $p_{jj} = 1$, then $\mathbb{P}(R_j = \infty \mid Z_0 = i) = p_{ij}$ by (1.25) and $\mathbb{E}[R_j \mid Z_0 = i] = \infty$.

$\square$

We check that if $p_{ij} = 0$ then $\mathbb{P}(R_j = 0 \mid Z_0 = i) = 1$, and $\mathbb{E}[R_j \mid Z_0 = i] = 0$.

1.4 Classification of States

This section presents the notions of communicating, transient and recurrent states, as well as the concept of irreducibility of a Markov chain. We also review the notions of positive and null recurrence, periodicity and aperiodicity of such chains. Those topics will be important when analysing the long-run behavior of Markov chains in the next chapter.

Communicating States

Definition 1.8 A state $(j) \in S$ is to be *accessible* from another state $(i) \in S$, and we write $(i) \longmapsto (j)$, if there exists a *finite* integer $n \geqslant 0$ such that

$$[P^n]_{i,j} = \mathbb{P}(Z_n = j \mid Z_0 = i) > 0.$$

In other words, it is possible to travel from (i) to (j) with non-zero probability in a certain (random) number of steps. We also say that state (i) leads to state (j), and when $i \neq j$ we have

$$\mathbb{P}(T_j^r < \infty \mid Z_0 = i) \geqslant \mathbb{P}(T_j^r \leqslant n \mid Z_0 = i) \geqslant \mathbb{P}(Z_n = j \mid Z_0 = i) > 0.$$

In case $(i) \longmapsto (j)$ and $(j) \longmapsto (i)$ we say that (i) and (j) *communicate* and we write $(i) \longleftrightarrow (j)$.

The binary relation "$\longleftrightarrow$" is a called an *equivalence relation* as it satisfies the following properties:

(a) *Reflexivity*:
   As $P^0 = I$ and $\mathbb{P}(Z_0 = i \mid Z_0 = i) = 1, i \in S$, for all $i \in S$ we have the relation $(i) \longleftrightarrow (i)$.

(b) *Symmetry*:
   For all $i, j \in S$ we have that $(i) \longleftrightarrow (j)$ is equivalent to $(j) \longleftrightarrow (i)$.

(c) *Transitivity*:

For all $i, j, k \in S$ such that $\textcircled{i} \longleftrightarrow \textcircled{j}$ and $\textcircled{j} \longleftrightarrow \textcircled{k}$, we have $\textcircled{i} \longleftrightarrow \textcircled{k}$.

Proof While points (a) and (b) are clearly valid, point (c) can be proved from the relation

$$\mathbb{P}(Z_{n+m} = k \mid Z_0 = i) = \sum_{l \in S} \mathbb{P}(Z_{n+m} = k \mid Z_n = l)\mathbb{P}(Z_n = l \mid Z_0 = i)$$

$$= \sum_{l \in S} \mathbb{P}(Z_m = k \mid Z_0 = l)\mathbb{P}(Z_n = l \mid Z_0 = i)$$

$$\geqslant \mathbb{P}(Z_m = k \mid Z_0 = j)\mathbb{P}(Z_n = j \mid Z_0 = i),$$

which shows that $\mathbb{P}(Z_{n+m} = k \mid Z_0 = i) > 0$ as soon as $\mathbb{P}(Z_m = k \mid Z_0 = j) > 0$ and $\mathbb{P}(Z_n = j \mid Z_0 = i) > 0$.   □

The equivalence relation '$\longleftrightarrow$" induces a *partition* of S into disjoint classes $A_1, A_2, \ldots, A_m, m \in \mathbb{N} \cup \{\infty\}$ such that $S = A_1 \cup \cdots \cup A_m$, and

(a) we have $\textcircled{i} \longleftrightarrow \textcircled{j}$ for all $i, j \in A_q$, and

(b) we have $\textcircled{i} \longleftrightarrow\!\!\!\!/ \ \textcircled{j}$ whenever $i \in A_p$ and $j \in A_q$ with $p \neq q$.

The sets $A_1, A_2, \ldots, A_m$ are called the *communicating classes* of the chain.

Definition 1.9 A Markov chain whose state space is made of a unique communicating class is said to be *irreducible*, otherwise the chain is said to be *reducible*.

The commands `communicatingClasses(MC)` and `is.irreducible(MC)` can be used together with the ⓡ code_02_page_16.R to determine the communicating classes and the irreducibility of a Markov chain.

Examples: Reducibility and Irreducibility

(i) Four communicating classes $\{0, 1\}$, $\{2\}$, $\{3\}$, and $\{4\}$:

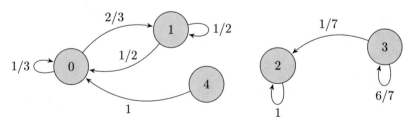

(ii) Two communicating classes {0, 1, 2} and {3}:

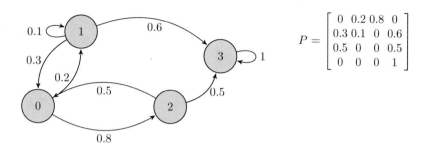

$$P = \begin{bmatrix} 0 & 0.2 & 0.8 & 0 \\ 0.3 & 0.1 & 0 & 0.6 \\ 0.5 & 0 & 0 & 0.5 \\ 0 & 0 & 0 & 1 \end{bmatrix}$$

(iii) Three communicating classes {0}, {1, 2}, {3, 4, 5}:

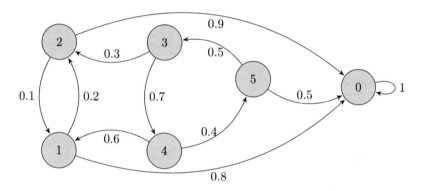

(iv) Two communicating classes {0} and {1, 2, 3}:

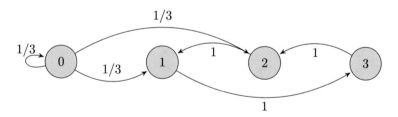

(v) Two communicating classes {0, 1, 2, 3} and {4}:

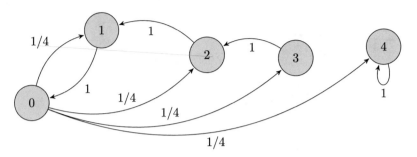

(vi) Five communicating classes {0}, {1}, {3}, {5}, and {2, 4}:

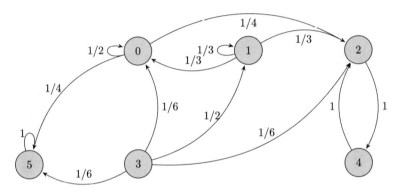

(vii) Three communicating classes {0, 2}, {1}, and {3}:

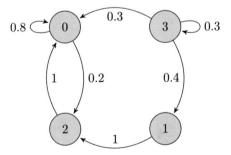

(viii) Two communicating classes $\{0, 1, 2\}$ and $\{3\}$:

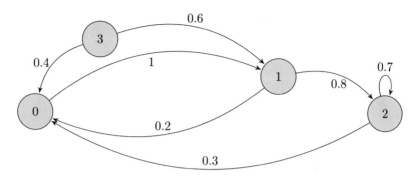

(ix) Four communicating classes $\{A, B\}, \{C\}, \{D\}, \{E\}$:

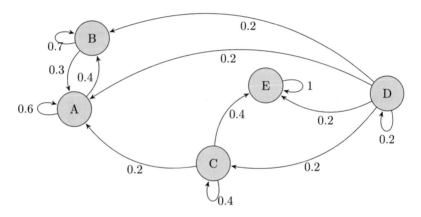

Recurrent States

Definition 1.10 A state $\textcircled{i} \in S$ is said to be *recurrent* if, starting from state $\textcircled{i}$, the chain will return to state $\textcircled{i}$ within a finite (random) time, with probability 1, i.e.,

$$p_{ii} := \mathbb{P}\big(T_i^r < \infty \mid Z_0 = i\big) = \mathbb{P}(Z_n = i \text{ for some } n \geqslant 1 \mid Z_0 = i) = 1. \tag{1.30}$$

The commands `recurrentStates(MC)` and `transientStates(MC)` can be used together with the ®R code_02_page_16.R to determine the recurrent and transient states of a Markov chain.

As a consequence of Propositions 1.5 and 1.7, the next result uses the mean number of returns R_i to state $\textcircled{i}$ defined in (1.23), and its proof relies on the geometric distribution (1.24) of R_i given that $Z_0 = i$. Note that the statements ((ii))–((iii)) below are not equivalent in general.

Proposition 1.11 *For any state* $(i) \in S$, *the following statements are equivalent:*

(i) the state $(i) \in S$ *is recurrent, i.e.* $p_{ii} = 1$,
(ii) the number of returns to $(i) \in S$ *is a.s.[2] infinite, i.e.*

$$\mathbb{P}(R_i = \infty \mid Z_0 = i) = 1, \ i.e. \ \mathbb{P}(R_i < \infty \mid Z_0 = i) = 0, \qquad (1.31)$$

(iii) the mean number of returns to $(i) \in S$ *is* infinite, *i.e.*

$$\mathbb{E}[R_i \mid Z_0 = i] = \infty, \qquad (1.32)$$

(iv) we have

$$\sum_{n \geqslant 1} f_{i,i}^{(n)} = 1, \qquad (1.33)$$

where $f_{i,i}^{(n)} := \mathbb{P}(T_i^r = n \mid Z_0 = i)$, $n \geqslant 1$, *is the distribution of* T_i^r.

As a consequence of Proposition 1.11, we also have the following characterization of recurrent states.

Corollary 1.12 *A state* $i \in S$ *is recurrent if and only if*

$$\sum_{n \geqslant 1} [P^n]_{i,i} = \infty,$$

i.e. the above series diverges.

Proof We have

$$\mathbb{E}[R_i \mid Z_0 = i] = \mathbb{E}\left[\sum_{n \geqslant 1} \mathbb{1}_{\{Z_n = i\}} \,\bigg|\, Z_0 = i\right]$$

$$= \sum_{n \geqslant 1} \mathbb{E}\left[\mathbb{1}_{\{Z_n = i\}} \,\big|\, Z_0 = i\right]$$

$$= \sum_{n \geqslant 1} \mathbb{P}(Z_n = i \mid Z_0 = i)$$

$$= \sum_{n \geqslant 1} [P^n]_{i,i},$$

and we conclude from Proposition 1.11.   □

---

[2] "Almost surely".

Corollary 1.12 admits the following consequence, which shows that any state communicating with a recurrent state is itself recurrent. In other words, recurrence is a *class property*, as all states in a given communicating class are recurrent as soon as one of them is recurrent, see, e.g., Corollary 6.6 in Privault (2018).

Corollary 1.13 (Class Property) *Let* $(j) \in S$ *be a recurrent state. Then any state* $(i) \in S$ *that communicates with state* (j) *is also recurrent.*

A communicating class $A \subset S$ is therefore recurrent if any of its states is recurrent.

Transient States

A state $(i) \in S$ is said to be *transient* when it is not recurrent, i.e., by (1.30),

$$p_{ii} = \mathbb{P}(T_i^r < \infty \mid Z_0 = i) = \mathbb{P}(Z_n = i \text{ for some } n \geqslant 1 \mid Z_0 = i) < 1, \tag{1.34}$$

or

$$\mathbb{P}(T_i^r = \infty \mid Z_0 = i) > 0.$$

Similarly to Proposition 1.11, we have the following result.

Proposition 1.14 *For any state* $(i) \in S$, *the following statements are equivalent:*

(i) the state $(i) \in S$ *is transient, i.e.* $p_{ii} < 1$,
(ii) the number of returns to $(i) \in S$ *is a.s.[3] finite, i.e.*

$$\mathbb{P}(R_i = \infty \mid Z_0 = i) = 0, \ i.e. \ \mathbb{P}(R_i < \infty \mid Z_0 = i) = 1, \tag{1.35}$$

(iii) the mean number of returns to $(i) \in S$ *is finite, i.e.*

$$\mathbb{E}[R_i \mid Z_0 = i] < \infty, \tag{1.36}$$

In other words, a state $(i) \in S$ is *transient* if and only if

$$\mathbb{P}(R_i < \infty \mid Z_0 = i) > 0,$$

which by (1.26) is equivalent to

$$\mathbb{P}(R_i < \infty \mid Z_0 = i) = 1,$$

---

[3] "Almost surely".

i.e. the number of returns to state $i \in S$ is finite with a non-zero probability which is necessarily equal to one. As a consequence of Corollary 1.12, we have the following result.

Corollary 1.15 *A state $i \in S$ is transient if and only if*

$$\sum_{n \geqslant 1} [P^n]_{i,i} < \infty,$$

i.e. the above series converges.

Similarly to Corollary 1.13, Corollary 1.15 admits the following consequence, which shows that any state communicating with a transient state is itself transient. Therefore, transience is also a *class property*, as all states in a given communicating class are transient as soon as one of them is transient.

Corollary 1.16 (Class Property) *Let $(j) \in S$ be a transient state. Then any state $(i) \in S$ that communicates with state (j) is also transient.*

Proof If a state $(i) \in S$ communicates with a transient state (j) then (i) is also transient (otherwise the state (j) would be recurrent by Corollary 1.13).   □

A communicating class $A \subset S$ is therefore transient if any of its states is transient.

Clearly, any absorbing state is recurrent, and any state that leads to an absorbing state is transient.

By analogy with (B.3), the matrix inverse

$$G := (I - P)^{-1} = \sum_{n \geqslant 0} P^n = I + \sum_{n \geqslant 1} P^n \tag{1.37}$$

of $I - P$ is called the *potential kernel*, or the *resolvent* of P, where I denotes the identity matrix.

Theorem 1.17 *Let $(Z_n)_{n \geqslant 0}$ be a Markov chain with* finite *state space* S. *Then $(Z_n)_{n \geqslant 0}$ has at least one recurrent state.*

Proof Corollary 1.15 and the relation

$$\sum_{n \geqslant 0} [P^n]_{i,j} = [(I - P)^{-1}]_{i,j}, \qquad i, j \in S, \tag{1.38}$$

show that a chain with finite state space is transient if the matrix $I - P$ is invertible. However, 0 is clearly an eigenvalue of $I - P$ with eigenvector $[1, 1, \ldots, 1]$, therefore $I - P$ is not invertible and as a consequence, a finite chain must admit at least one recurrent state. □

The next proposition is applied to the Snakes and Ladders game in e.g. .

Proposition 1.18 *Assume that the chain $(Z_n)_{n \geq 0}$ has a* finite *state space* $S = \{1, \ldots, m\}$ *made of* $\{1, \ldots, m-1\}$ *transient states and a* unique *absorbing state* $\boxed{m}$. *Then, we have the expression*

$$\mathbb{E}[T_m \mid Z_0 = i] = \sum_{\substack{j \in S \\ j \neq m}} \left[[I - Q]^{-1}\right]_{i,j}, \quad i \neq m, \tag{1.39}$$

where Q is the matrix $Q := (P_{i,j})_{1 \leq i,j \leq m-1}$.

Proof By (1.23), since the states $\{1, \ldots, m-1\}$ are transient, we have

$$\mathbb{E}[R_j \mid Z_0 = i] = \mathbb{E}\left[\sum_{n \geq 1} \mathbb{1}_{\{Z_n = j\}} \,\middle|\, Z_0 = i\right]$$

$$= \sum_{n \geq 1} \mathbb{E}[\mathbb{1}_{\{Z_n = j\}} \mid Z_0 = i]$$

$$= \sum_{n \geq 1} \mathbb{P}(Z_n = j \mid Z_0 = i)$$

$$= \sum_{n \geq 1} [Q^n]_{i,j}$$

$$= -\mathbb{1}_{\{i=j\}} + \sum_{n \geq 0} [Q^n]_{i,j}$$

$$< \infty, \quad 1 \leq i, j \leq m - 1.$$

Hence Q is invertible, and we have

$$\mathbb{E}[R_j \mid Z_0 = i] = -\mathbb{1}_{\{i=j\}} + \left[[I - Q]^{-1}\right]_{i,j}, \quad 1 \leq i, j \leq m - 1.$$

On the other hand, after starting from $i \in \{1, \ldots, m-1\}$, we have

$$T_m = 1 + \sum_{\substack{j \in S \\ j \neq m}} R_j,$$

hence

$$\mathbb{E}[T_m \mid Z_0 = i] = 1 + \sum_{\substack{j \in S \\ j \neq m}} \mathbb{E}[R_j \mid Z_0 = i]$$

$$= 1 + \sum_{\substack{j \in S \\ j \neq m}} \left(-\mathbb{1}_{\{i=j\}} + \left[[I - Q]^{-1} \right]_{i,j} \right)$$

$$= \sum_{\substack{j \in S \\ j \neq m}} [[I - Q]^{-1}]_{i,j}.$$

□

Examples: Recurrent and Transient States

(i) States $①$, $②$, $③$, $④$ and $⑤$ are transient, and state $⓪$ is recurrent.

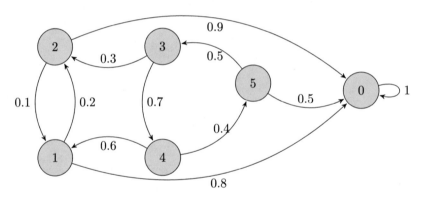

(ii) State $⓪$ is transient, and states $①$, $②$, $③$ are recurrent.

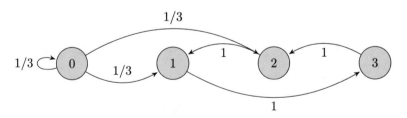

(iii) State ④ is absorbing (and therefore recurrent), state ⓪ is transient and the remaining states ①, ②, ③ are also transient because they communicate with the transient state ⓪.

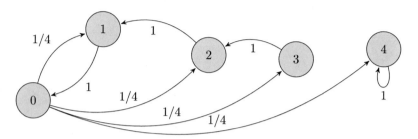

(iv) States ⓪, ①, ③ are transient and states ②, ④, ⑤ are recurrent.

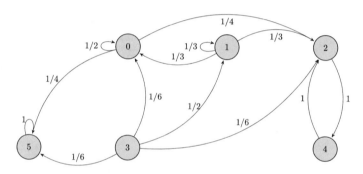

(v) States ① and ③ are transient, states ⓪ and ② are recurrent by Proposition 1.13 and Theorem 1.17.

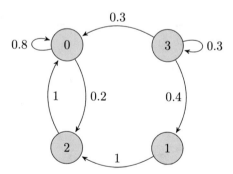

(vi) State ③ is transient, and states ⓪,①,② are recurrent.

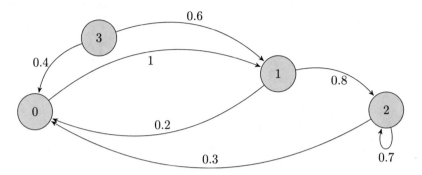

(vii) States Ⓐ, Ⓑ and Ⓔ are recurrent, and states Ⓒ, Ⓓ are transient.

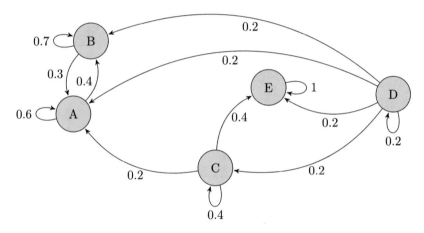

(viii) States ③ and ④ are transient, states ⓪ and ① are recurrent, and state ②
is absorbing (hence it is recurrent).

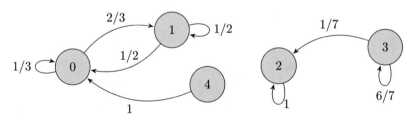

Positive vs. Null Recurrence

The expected time of return (or mean recurrence time) to a state $(i) \in S$ is given by

$$\mu_i(i) := \mathbb{E}\big[T_i^r \,\big|\, Z_0 = i\big] = \sum_{n \geqslant 1} n \mathbb{P}\big(T_i^r = n \,\big|\, Z_0 = i\big).$$

Recall that a state (i) is recurrent when $\mathbb{P}\big(T_i^r < \infty \,\big|\, Z_0 = i\big) = 1$, i.e. when the random return time T_i^r is almost surely *finite* starting from state (i). However, the recurrence property yields no information on the finiteness of its expectation $\mu_i(i) = \mathbb{E}\big[T_i^r \,\big|\, Z_0 = i\big]$, $i \in S$.

Definition 1.19 A *recurrent* state $i \in S$ is said to be:

(a) *positive recurrent* if the mean return time to (i) is *finite*, i.e.

$$\mu_i(i) = \mathbb{E}\big[T_i^r \,\big|\, Z_0 = i\big] < \infty,$$

(b) *null recurrent* if the mean return time to (i) is *infinite*, i.e.

$$\mu_i(i) = \mathbb{E}\big[T_i^r \,\big|\, Z_0 = i\big] = \infty.$$

The following Theorem 1.20 shows in particular that a Markov chain with finite state space cannot have any null recurrent state, cf. e.g. Corollary 2.3 in Kijima (1997), and also Corollary 3.7 in .

Theorem 1.20 *Assume that the state space* S *of a Markov chain* $(Z_n)_{n \geqslant 0}$ *is finite. Then, any recurrent state in* S *is also positive recurrent.*

As a consequence of Definition 1.9, Corollary 1.13, and Theorems 1.17 and 1.20 we have the following corollary.

Corollary 1.21 *Let* $(Z_n)_{n \geqslant 0}$ *be an irreducible Markov chain with* finite *state space* S. *Then all states of* $(Z_n)_{n \geqslant 0}$ *are positive recurrent.*

Periodicity and Aperiodicity

Given a state $i \in S$, consider the sequence

$$\{n \geqslant 1 \; : \; [P^n]_{i,i} > 0\}$$

of integers which represent the possible travel times from state (i) to itself.

Definition 1.22 The *period* of the state $i \in S$ is the greatest common divisor of the sequence

$$\{n \geqslant 1 \,:\, [P^n]_{i,i} > 0\}.$$

A state $i \in S$ having period 1 is said to be *aperiodic*. This is the case in particular when $P_{i,i} > 0$, i.e. when (i) admits a returning loop with nonzero probability.

In particular, any absorbing state is both aperiodic and recurrent. A *recurrent* state $i \in S$ is said to be *ergodic* if it is both *positive recurrent* and *aperiodic*.

If $[P^n]_{i,i} = 0$ for all $n \geqslant 1$ then the set $\{n \geqslant 1 \,:\, [P^n]_{i,i} > 0\}$ is empty and by convention the period of state (i) is defined to be 0. In this case, state (i) is also transient.

Note also that if

$$\{n \geqslant 1 \,:\, [P^n]_{i,i} > 0\}$$

contains two distinct numbers that are relatively prime to each other (i.e. their greatest common divisor is 1) then state (i) aperiodic.

Proposition 1.23 shows that periodicity is a *class property*, as all states in a given communicating class have the same periodicity, see, e.g., Corollary 6.14 in Privault (2018).

Proposition 1.23 (Class Property) *All states that belong to the same communicating class have the same period.*

A Markov chain is said to be *aperiodic* when *all* of its states are aperiodic. Note that any state that communicates with an aperiodic state becomes itself aperiodic. In particular, if a communicating class contains an aperiodic state then the whole class becomes aperiodic.

The command `period(MC)` in the ®️ markovchain package can be used together with the ®️ code_02_page_16.R to determine the periodicity of an irreducible Markov chain.

Examples: Periodicity and Aperiodicity

(i) All states have period 4.

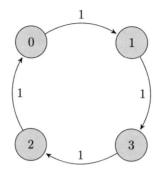

(ii) All states have period 2.

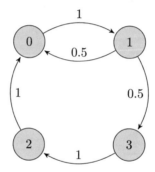

(iii) All states have period 1.

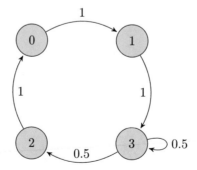

(1.40)

(iv) All states have period 1.

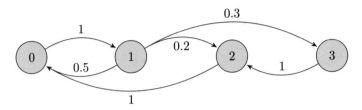

(v) All states have period 1.

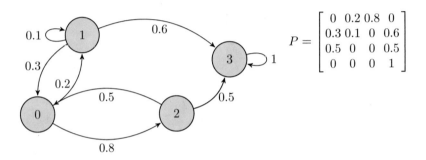

$$P = \begin{bmatrix} 0 & 0.2 & 0.8 & 0 \\ 0.3 & 0.1 & 0 & 0.6 \\ 0.5 & 0 & 0 & 0.5 \\ 0 & 0 & 0 & 1 \end{bmatrix}$$

(vi) State ⓪ has period 1, states ① and ② have period 2, and states ③, ④ and ⑤ have period 3.

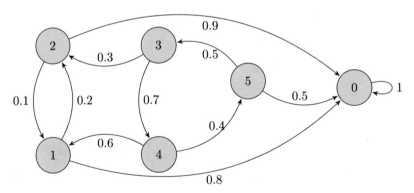

(vii) State ⓪ has period 1 and states ①, ②, ③ have period 3.

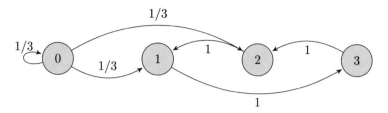

(viii) All states have period 1.

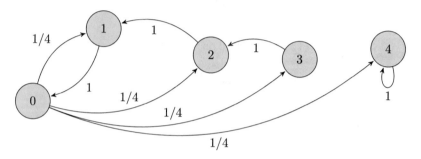

(ix) State ③ has period 0, states ② and ④ have period 2, and states ⓪, ①, ⑤ are aperiodic.

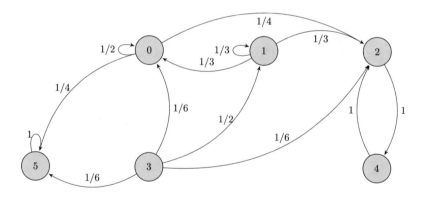

(x) States ⓪, ②, ③ have period 1, and state ① has period 0.

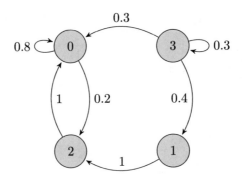

(xi) States ⓪, ①, ②, have period one, and state ③ has period 0.

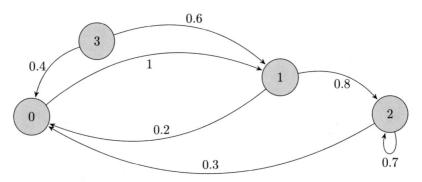

(xii) States ⓪, ①, ②, ③ have period one, and state ④ has period 0.

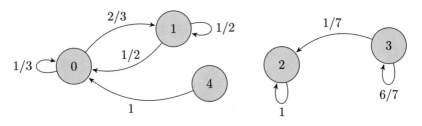

1.5 Hitting Times of Random Walks

This section reviews some basic results on the hitting times of the one-dimensional random walk $(S_n)_{n \geqslant 0}$, defined by $S_0 = 0$ and

$$S_n = \sum_{k=1}^{n} X_k = X_1 + \cdots + X_n, \qquad n \geqslant 0.$$

Here, the random walk *increments*

$$X_n \in \{-1, +1\}, \qquad n \geqslant 1,$$

form an *independent and identically distributed (i.i.d.)* family of Bernoulli random variables, with distribution

$$\begin{cases} \mathbb{P}(X_k = +1) = p, \\ \\ \mathbb{P}(X_k = -1) = q, \qquad k \geqslant 1, \end{cases}$$

with $p + q = 1$. This one-dimensional random walk can only evolve by going up of down by one unit within the finite state space $S = \{0, 1, \ldots, L\}$. We have

$$\mathbb{P}(S_{n+1} = k + 1 \mid S_n = k) = p \text{ and } \mathbb{P}(S_{n+1} = k - 1 \mid S_n = k) = q,$$

$k \in \mathbb{Z}$. We also have

$$\mathbb{E}[S_n \mid S_0 = 0] = \mathbb{E}\left[\sum_{k=1}^{n} X_k\right] = \sum_{k=1}^{n} \mathbb{E}[X_k] = n(2p - 1) = n(p - q),$$

and the variance can be computed as

$$\mathrm{Var}[S_n \mid S_0 = 0] = \mathrm{Var}\left[\sum_{k=1}^{n} X_k\right] = \sum_{k=1}^{n} \mathrm{Var}[X_k] = 4npq.$$

Let

$$T_L := \inf\{n \geqslant 0 \ : \ S_n = L\}$$

Fig. 1.1 Sample path of the
random walk $(S_n)_{n\geqslant 0}$

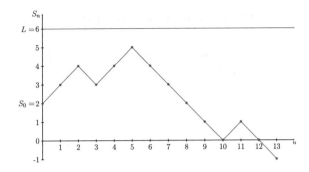

denote the first hitting time of L by the one-dimensional random walk $(S_n)_{n\geqslant 0}$, and
let

$$T_0 := \inf\{n \geqslant 0 \ : \ S_n = 0\}$$

denote the first hitting time of 0 by the process $(S_n)_{n\geqslant 0}$ (Fig. 1.1).
See e.g. Relation (2.2.27) in Privault (2018) for the following proposition.

Proposition 1.24 *In the non-symmetric case $p \neq q$, the event*

$$\{T_0 < T_L\} = \bigcup_{n\geqslant 0}\{S_n = 0\}, \tag{1.41}$$

has the conditional probability

$$\mathbb{P}(T_0 < T_L \mid S_0 = k) = \frac{(q/p)^k - (q/p)^L}{1 - (q/p)^L} = \frac{(p/q)^{L-k} - 1}{(p/q)^L - 1}, \tag{1.42}$$

or

$$\mathbb{P}(T_L < T_0 \mid S_0 = k) = \frac{(p/q)^{L-k} - (p/q)^L}{1 - (p/q)^L} = \frac{1 - (q/p)^k}{1 - (q/p)^L}, \tag{1.43}$$

$k = 0, 1, \ldots, L$.

In the symmetric case $p = q = 1/2$, we find

$$\mathbb{P}(T_0 < T_L \mid S_0 = k) = 1 - \frac{k}{L}, \quad \text{or} \quad \mathbb{P}(T_L < T_0 \mid S_0 = k) = \frac{k}{L}, \tag{1.44}$$

$k = 0, 1, \ldots, L$, see Relation (2.2.28) in Privault (2018). When the number L of
states becomes large we obtain the probability of hitting the origin starting from
state $\textcircled{k}$, as

$$f_\infty(k) := \mathbb{P}(T_0 < \infty \mid S_0 = k)$$

$$= \mathbb{P}\left(\bigcup_{L \geqslant 1} \{T_0 < T_L\} \,\Big|\, S_0 = k \right) \tag{1.45}$$

$$= \min\left(1, \left(\frac{q}{p}\right)^k \right)$$

$$= \begin{cases} 1 & \text{if } q \geqslant p, \\ \left(\dfrac{q}{p}\right)^k & \text{if } p > q, \quad k \geqslant 0. \end{cases} \tag{1.46}$$

Similarly, for all $k \geqslant 0$ we have

$$\mathbb{P}(T_0 = \infty \mid S_0 = k) = \mathbb{P}\left(\bigcap_{L \geqslant 1} \{T_L < T_0\} \,\Big|\, S_0 = k \right)$$

$$= \lim_{L \to \infty} \mathbb{P}(T_L < T_0 \mid S_0 = k)$$

$$= \begin{cases} 0 & \text{if } p \leqslant q, \\ 1 - \left(\dfrac{q}{p}\right)^k & \text{if } p > q, \end{cases}$$

which represents the probability that the one-dimensional random walk $(S_n)_{n \geqslant 0}$ "escapes to infinity".

Mean Hitting Times

Let now

$$T_{0,L} = \inf\{n \geqslant 0 \ : \ S_n = 0 \text{ or } S_n = L\} \tag{1.47}$$

denote the time[4] until any of the states $\textcircled{0}$ or $\textcircled{L}$ is reached by $(S_n)_{n \geqslant 0}$, with $T_{0,L} = +\infty$ in case neither states are ever reached, see Fig. 1.2.

---

[4] The notation "inf" stands for "infimum", meaning the smallest $n \geqslant 0$ such that $S_n = 0$ or $S_n = L$, if such an n exists.

Fig. 1.2 Sample paths of the random walk $(S_n)_{n \geqslant 0}$

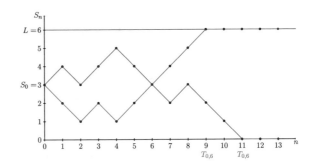

From Proposition 1.24, we note that

$$
\mathbb{P}(T_0 < T_L \mid S_0 = k) + \mathbb{P}(T_L < T_0 \mid S_0 = k)
$$

$$
= \frac{(p/q)^{L-k} - 1}{(p/q)^L - 1} + \frac{(q/p)^k - 1}{(q/p)^L - 1}
$$

$$
= \frac{(q/p)^L((p/q)^{L-k} - 1) - ((p/q)^{L-k} - 1) + (p/q)^L((q/p)^k - 1) - ((q/p)^k - 1)}{((p/q)^L - 1)((q/p)^L - 1)}
$$

$$
= \frac{(q/p)^k - (q/p)^L - (p/q)^{L-k} + 1 + (p/q)^{L-k} - (p/q)^L - (q/p)^k + 1}{((p/q)^L - 1)((q/p)^L - 1)}
$$

$$
= 1, \qquad k = 0, 1, \ldots, L, \tag{1.48}
$$

see Exercise 1.2.

We refer to Relation (2.3.11) in Privault (2018) for the following proposition.

Proposition 1.25 *When $p \neq q$, the mean hitting time*

$$
h_L(k) := \mathbb{E}[T_{0,L} \mid S_0 = k]
$$

starting from $S_0 = k \in \{0, 1, \ldots, L\}$ can be computed as

$$
h_L(k) = \mathbb{E}[T_{0,L} \mid S_0 = k] = \frac{1}{q - p}\left(k - L\frac{1 - (q/p)^k}{1 - (q/p)^L}\right), \qquad k = 0, 1, 2, \ldots, L. \tag{1.49}
$$

In the symmetric case $p = q = 1/2$, we get

$$
h_L(k) = \mathbb{E}[T_{0,L} \mid S_0 = k] = k(L - k), \qquad k = 0, 1, 2, \ldots, L, \tag{1.50}
$$

see Relation (2.3.17) in Privault (2018). In particular, we note that

$$
\mathbb{E}[T_{0,L} \mid S_0 = k] < +\infty, \qquad k = 0, 1, 2, \ldots, L.
$$

Notes

See e.g. Chen and Hong (2012) for statistical testing of the Markov property in time series, and Billingsley (1961), Azais and Bouguet (2018), and Broemeling (2018), for references on statistical inference for Markov chains. Additional background on the Markov property can be found in Chapters 4–6 in Privault (2018), and in references therein.

Exercises

Exercise 1.1 Consider the Markov chain $(X_n)_{n \geqslant 0}$ on $S = \{0, 1, 2\}$ whose transition probability matrix P is given by

$$P = \begin{array}{c} \\ 0 \\ 1 \\ 2 \end{array} \begin{array}{ccc} 0 & 1 & 2 \\ \left[\begin{array}{ccc} 1 & 0 & 0 \\ 1/4 & 0 & 3/4 \\ 0 & 1 & 0 \end{array} \right] \end{array}.$$

(a) Draw a graph of the chain and find the probability $g_0(k)$ that the chain is absorbed into state ⓪ given that it started from states $k = 0, 1, 2$.
(b) Determine the mean time $h_0(k)$ it takes until the chain is absorbed into state ⓪, after starting from $k = 0, 1, 2$.

Exercise 1.2 Recover Relation (1.48) by showing independently that for all $k = 0, 1, \ldots, L$ we have $\mathbb{P}(T_{0,L} < \infty \mid S_0 = k) = 1$, i.e. the stopping time $T_{0,L}$ defined in (1.47) is finite almost surely.

Exercise 1.3 Consider the Markov chain $(X_n)_{n \geqslant 0}$ with state space $S = \{0, 1, 2, 3\}$ and transition probability matrix given by

$$[P_{i,j}]_{0 \leqslant i,j \leqslant 3} = \begin{bmatrix} 1 & 0 & 0 & 0 \\ 0.3 & 0 & 0.4 & 0.3 \\ 0.3 & 0.4 & 0 & 0.3 \\ 0 & 0 & 0 & 1 \end{bmatrix}.$$

(a) What are the absorbing states of the chain $(X_n)_{n \geqslant 0}$?
(b) Denoting by $T_k := \inf\{n \geqslant 0 : X_n = k\}$ the first hitting time of state ⓚ, find the probabilities $g_1(k) = \mathbb{P}(T_1 < \infty \mid X_0 = k)$ of hitting state ① in finite time after starting from state ⓚ, for $k = 0, 1, 2, 3$.

(c) Denoting by $T_1^r := \inf\{n \geq 1 : X_n = 1\}$ the first *return time* to state (1), find the probabilities $p_1(k) = \mathbb{P}(T_1^r < \infty \mid X_0 = k)$ of returning to state (1) in finite time after starting from state (k), for $k = 0, 1, 2, 3$.

(d) Find the mean hitting times $h_1(k) = \mathbb{E}[T_1 \mid X_0 = k]$ of state (1) and the mean return times $\mu_1(k) = \mathbb{E}[T_1 \mid X_0 = k]$ to state (1) after starting from state (k), for $k = 0, 1, 2, 3$.

Exercise 1.4 A box contains red balls and green balls. At each time step we pick a ball uniformly at random and without replacement. If the ball is red we lose $1, and if the ball is green we gain +$1. The game ends when the box becomes empty. We let $f(x, y)$ denote the value of the game when the game starts with $x \geq 0$ red balls and $y \geq 0$ green balls in the box.

(a) Find the boundary conditions $f(x, 0)$, $x \geq 0$, and $f(0, y)$, $y \geq 0$.
(a) Using first step analysis, derive the finite difference equation satisfied by $f(x, y)$ for $x, y \geq 1$.
(b) Solve the equation of Question (1.4) for $f(x, y)$, $x, y = 1, 2, 3$.
(c) Find $f(x, y)$ for all $x, y \geq 0$.

Exercise 1.5 Two buffalos are traveling in opposite directions on a one-dimensional road $\{0, 1, \ldots, S\}$, one step at a time. Buffalo A starts from (0), moving up by +1 at every time step, and Buffalo B starts at the same time from (S), moving down by -1 at every time step.

(a) How many time steps does it take for Buffalo A to travel up from (0) to (S), and for Buffalo B to travel down from (S) to (0)?
(b) Next, we assume that when the buffalos collide, they either both continue the same ways with probability p, or they both turn back and continue in opposite directions with probability $q = 1 - p$. How many time steps does it take for the buffalos to reach any of the boundaries (0) or (S)?

Exercise 1.6 Consider the Markov chain $(X_n)_{n \geq 0}$ on the countably infinite state space $S = \mathbb{N} = \{0, 1, 2, 3, \ldots\}$, with the infinite transition matrix

$$P = \left[\, P_{i,j} \,\right]_{i,j \in \mathbb{N}} = \begin{bmatrix} q & p & 0 & 0 & 0 & \cdots \\ 0 & q & p & 0 & 0 & \cdots \\ 0 & 0 & q & p & 0 & \cdots \\ 0 & 0 & 0 & q & p & \cdots \\ \vdots & \vdots & \vdots & \vdots & \vdots & \ddots \end{bmatrix},$$

where $p, q \in (0, 1)$ are such that $p + q = 1$.

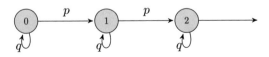

(a) By a recurrence using Pascal's identity

$$\binom{n}{k} = \binom{n-1}{k-1} + \binom{n-1}{k},$$

  compute $[P^n]_{i,j}$, $n \geqslant 1$, in the cases (1) $j - i \leqslant n$, (2) $n < j - i$, (3) $i > j$.
(b) Show that for all $i, j \geqslant 0$ we have

$$\lim_{n \to \infty} [P^n]_{i,j} = 0.$$

(c) Compute

$$\sum_{n \geqslant 0} [P^n]_{i,j}$$

  in the cases (1) $i \leqslant j$, (2) $i > j$.
(d) Letting $T_j := \inf\{n \geqslant 0 : X_n = j\}$, determine the value of

$$p_{i,j} := \mathbb{P}(T_j < \infty \mid X_0 = i)$$

  in the cases (1) $i < j$, (2) $i = j$, (3) $i > j$.
(e) Is the chain $(X_n)_{n \geqslant 0}$ recurrent or transient?
(f) Compute the mean number of returns $\mathbb{E}[R_j \mid X_0 = i]$ from state $\widehat{i}$ to state $\widehat{j}$
  in the cases (1) $i < j$, (2) $i = j$, (3) $i > j$.
(g) Show that the matrix $I - P$ is invertible, and compute its inverse $(I - P)^{-1}$.

Exercise 1.7 Ring toss game. Let $\mathbb{N} := \{0, 1, 2, \ldots\}$ and consider the two-dimensional random walk $(Z_k)_{k \in \mathbb{N}} = (X_k, Y_k)_{k \in \mathbb{N}}$ on $\mathbb{N} \times \mathbb{N}$ with the transition probabilities

$$\mathbb{P}\big((X_{k+1}, Y_{k+1}) = (x + 1, y) \mid (X_k, Y_k) = (x, y)\big)$$

$$= \mathbb{P}\big((X_{k+1}, Y_{k+1}) = (x, y + 1) \mid (X_k, Y_k) = (x, y)\big) = \frac{1}{2}, \quad (x, y) \in \mathbb{N} \times \mathbb{N},$$

$k \geqslant 0$, and let

$$A := \mathbb{N}^2 \setminus \{0, 1, 2\}^2 = \big\{(x, y) \in \mathbb{N} \times \mathbb{N} : x \geqslant 3 \text{ or } y \geqslant 3\big\}.$$

Let also

$$T_A := \inf\big\{n \geqslant 0 : (X_n, Y_n) \in A\big\}$$

Table 1.1 Domain A with $N = 3$ (in blue)

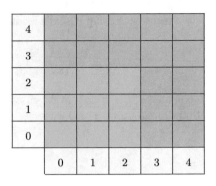

denote the first hitting time of the set A by the random walk $(Z_k)_{k\in\mathbb{N}} = (X_k, Y_k)_{k\in\mathbb{N}}$, see Table 1.1, and consider the mean hitting times

$$\mu_A(x, y) := \mathbb{E}\big[T_A \mid (X_0, Y_0) = (x, y)\big], \quad (x, y) \in \mathbb{N} \times \mathbb{N}.$$

(a) Give the values of $\mu_A(x, y)$ when $(x, y) \in A$.
(b) By applying first step analysis, find an equation satisfied by $\mu_A(x, y)$ on the domain

$$A^c = \big\{(x, y) \in \mathbb{N} \times \mathbb{N} : 0 \leqslant x, y \leqslant 3\big\}.$$

(c) Find the values of $\mu_A(x, y)$ for all $x, y \leqslant 3$ by solving the equation of Question (1.7).
(d) In each round of a ring toss game, a ring is thrown at two sticks in such a way that each stick has exactly 50% chance to receive the ring. Compute the mean time it takes until at least one of the two sticks receives three rings.

Exercise 1.8 Taking $\mathbb{N} := \{0, 1, 2, \ldots\}$, consider the two-dimensional random walk $(Z_k)_{k\in\mathbb{N}} = (X_k, Y_k)_{k\in\mathbb{N}}$ on $\mathbb{N} \times \mathbb{N}$ with the transition probabilities

$$\mathbb{P}(X_{k+1} = x + 1, \ Y_{k+1} = y \mid X_k = x, \ Y_k = y)$$
$$= \mathbb{P}(X_{k+1} = x, \ Y_{k+1} = y + 1 \mid X_k = x, \ Y_k = y)$$
$$= \frac{1}{2}, \quad k \geqslant 0,$$

and let

$$A = [2, \infty) \times [2, \infty) = \big\{(x, y) \in \mathbb{N} \times \mathbb{N} : x \geqslant 2, \ y \geqslant 2\big\}.$$

Let also

$$T_A := \inf\{n \geqslant 0 : X_n \geqslant 2 \text{ and } Y_n \geqslant 2\}$$

Table 1.2 Domain A with $N = 2$ (in blue)

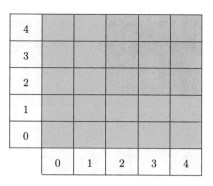

denote the hitting time of the set A by the random walk $(Z_k)_{k \in \mathbb{N}}$, see Table 1.2, and consider the mean hitting times

$$\mu_A(x, y) := \mathbb{E}[T_A \mid X_0 = x, \ Y_0 = y], \qquad x, y \in \mathbb{N}.$$

(a) Give the value of $\mu_A(x, y)$ when $x \geqslant 2$ and $y \geqslant 2$.

(b) Show that $\mu_A(x, y)$ solves the equation

$$\mu_A(x, y) = 1 + \frac{1}{2}\mu_A(x+1, y) + \frac{1}{2}\mu_A(x, y+1), \qquad x, y \in \mathbb{N}. \tag{1.51}$$

(c) Show that $\mu_A(1, 2) = \mu_A(2, 1) = 2$ and $\mu_A(0, 2) = \mu_A(2, 0) = 4$.

(d) In each round of a ring toss game, a ring is thrown at two sticks in such a way that each stick has exactly 50% chance to receive the ring. Compute the mean time it takes until both sticks receive at least two rings.

Problem 1.9 (Chen (2004), Propositions 2.14–2.15)

Given $(X_n)_{n \geqslant 0}$ a Markov chain with transition probability matrix $P = (P_{i,j})_{i,j \in S}$ on a state space S and $v = (v_k)_{k \in S}$ a nonnegative vector, we say that $u^* = (u_i^*)_{i \in S}$ is the minimal non-negative solution to the equation

$$u_i = v_i + \sum_{k \in S} P_{i,k} u_k, \qquad i \in S. \tag{1.52}$$

if u^* satisfies (1.52) and any other solution u of (1.52) satisfies $u_i \geqslant u_i^*, i \in S$. For $i, j \in S$, let

$$f_{i,j}^{(n)} = \mathbb{P}(X_n = j, X_{n-1} \neq j, \ldots, X_1 \neq j \mid X_0 = i) = \mathbb{P}(T_j = n \mid X_0 = i),$$

$n \geqslant 0$, where T_j denotes the first hitting time of state $\textcircled{j}$ by $(X_n)_{n \geqslant 0}$.

(a) Give the value of $f_{i,j}^{(1)}$ from the transition probability matrix P.

(b) Using first step analysis, show that for all $j \in S$, $\left(f_{i,j}^{(n)}\right)_{i \in S}$ satisfies the equation

$$f_{i,j}^{(n+1)} = \sum_{\substack{k \in S \\ k \neq j}} P_{i,k} f_{k,j}^{(n)}, \qquad i, j \in S, \ n \geqslant 0. \tag{1.53}$$

(c) Let

$$f_{i,j} := \mathbb{P}(T_j < \infty \mid X_0 = i) = \sum_{n \geqslant 1} f_{i,j}^{(n)}, \qquad i, j \in S.$$

Show that

$$f_{i,j} = P_{i,j} + \sum_{\substack{k \in S \\ k \neq j}} P_{i,k} f_{k,j}, \qquad i, j \in S. \tag{1.54}$$

(d) Show that for all $j \in S$, $(f_{i,j})_{i \in S}$ is the unique minimal solution to Eq. (1.54).
    Hint: Letting $\tilde{f}$ denote another solution of (1.54), show, using (1.53) and induction on $n \geqslant 1$, that

$$\tilde{f}_{i,j} \geqslant \sum_{l=1}^{n} f_{i,j}^{(l)}, \quad i, j \in S, \quad n \geqslant 1.$$

(e) Let $g_{i,j}^{(1)} := f_{i,j}^{(1)}$ and

$$g_{i,j}^{(n+1)} := f_{i,j}^{(n+1)} + n \sum_{\substack{k \in S \\ k \neq j}} P_{i,k} f_{k,j}^{(n)}, \quad i, j \in S, \quad n \geqslant 1.$$

Using (1.53), show by induction on n that $g_{i,j}^{(n)} = n f_{i,j}^{(n)}$, $i, j \in S, n \geqslant 1$.

(f) Let

$$h_{i,j} := \mathbb{E}[T_j < \infty \mid X_0 = i] = \sum_{n \geqslant 1} n \mathbb{P}(T_j = n \mid X_0 = i)$$

$$= \sum_{n \geqslant 1} g_{i,j}^{(n)}, \quad i, j \in S.$$

Show that

$$h_{i,j} = f_{i,j} + \sum_{\substack{k \in S \\ k \neq j}} P_{i,k} h_{k,j}, \qquad i, j \in S, \tag{1.55}$$

where

$$f_{i,j} := \mathbb{P}(T_j < \infty \mid X_0 = i) = \sum_{n \geqslant 1} f_{i,j}^{(n)}, \qquad i, j \in S.$$

(g) Show that for all $j \in S$, $(h_{i,j})_{i \in S}$ is the unique minimal solution to Eq. (1.55). *Hint:* Letting $\tilde{h}$ denote another solution of (1.55), show, using (1.53) and induction on $n \geqslant 1$, that

$$\tilde{h}_{i,j} \geqslant \sum_{l=1}^{n} g_{i,j}^{(l)}, \qquad i, j \in S, \quad n \geqslant 1.$$

Chapter 2
Phase-Type Distributions

Phase-type distributions (Neuts (1981)) provide a class of probability distributions depending on a flexible range of parameters, that can be used to fit actual data. Phase-type distributions are used for modeling and simulation in insurance, risk management and actuarial science, where they can be used to model heavy-tailed random claim sizes appearing for example in reserve and surplus processes.

2.1 Negative Binomial Distribution

Given $p \in [0, 1]$, consider a two-state Markov chain $(X_n)_{n \geqslant 0}$ on the state space $\{0, 1\}$, with transition matrix

$$P = \begin{bmatrix} 1 & 0 \\ q & p \end{bmatrix},$$

with $q := 1 - p$. We note that

(i) State $\textcircled{0}$ is absorbing, i.e. $\mathbb{P}(X_{n+1} = 0 \mid X_n = 0) = 1$, *and*
(ii) The first hitting time

$$T_0 := \inf\{n \geqslant 0 \; : \; X_n = 0\}$$

of state $\textcircled{0}$ starting from state $\textcircled{1}$ has the geometric distribution p given by

$$\mathbb{P}(T_0 = k \mid X_0 = 1) = (1 - p)p^{k-1}, \qquad k \geqslant 1.$$

© The Author(s), under exclusive license to Springer Nature Switzerland AG 2024
N. Privault, *Discrete Stochastic Processes*, Springer Undergraduate
Mathematics Series, https://doi.org/10.1007/978-3-031-65820-4_2

More generally, given $d \geqslant 1$, consider a $d + 1$-state Markov chain $(X_n)_{n \geqslant 0}$ on the state space $\{0, 1, \ldots, d\}$, with transition matrix

$$P = \begin{bmatrix} 1 & 0 & 0 & \cdots & 0 & 0 \\ q & p & 0 & \cdots & 0 & 0 \\ 0 & q & p & \cdots & 0 & 0 \\ \vdots & \vdots & \ddots & \ddots & \vdots & \vdots \\ 0 & 0 & \cdots & q & p & 0 \\ 0 & 0 & \cdots & 0 & q & p \end{bmatrix},$$

with $q := 1 - p$. In this case,

(i) State $\textcircled{0}$ is absorbing, i.e. $\mathbb{P}(X_{k+1} = 0 \mid X_k = 0) = 1$, and
(ii) The first hitting time T_0 of state $\textcircled{0}$ starting from state $\textcircled{d}$ has the (shifted) negative binomial distribution

$$\mathbb{P}(T_0 = k \mid X_0 = d) = \binom{k-1}{k-d}(1-p)^d p^{k-d}, \qquad k \geqslant d.$$

2.2   Markovian Construction

The idea of phase-type distributions is to generalize the above modeling by considering a discrete-time Markov chain $(X_n)_{n \geqslant 0}$ on $\{0, 1, \ldots, d\}$ having d transient[1] states $\{1, 2, \ldots, d\}$, and $\textcircled{0}$ as absorbing state. The geometric and negative binomial distributions have power tails, hence they are examples of heavy-tailed probability distributions.

Clearly, the first row of P has to be $[1, 0, \ldots, 0]$ because state $\textcircled{0}$ is absorbing, and the remaining of the matrix can take the form $[\alpha, Q]$. Hence the transition matrix P of the chain $(X_n)_{n \geqslant 0}$ takes the form

$$P = \big[P_{i,j}\big]_{0 \leqslant i,j \leqslant d} = \begin{bmatrix} 1 & 0 & \cdots & 0 \\ \alpha_1 & Q_{1,1} & \cdots & Q_{1,d} \\ \alpha_2 & Q_{2,1} & \cdots & Q_{2,d} \\ \vdots & \vdots & \ddots & \vdots \\ \alpha_d & Q_{d,1} & \cdots & Q_{d,d} \end{bmatrix} = \begin{bmatrix} 1 & 0 \\ \alpha & Q \end{bmatrix},$$

---

[1] Here, the transience condition implies that $\mathbb{P}(T_0 < \infty \mid X_0 = i) = 1$ for all $i = 1, 2, \ldots, d$, it will be ensured by assuming that $I - Q$ is invertible, see Sect. 1.4 for details.

where α is the column vector $\alpha = [\alpha_1, \alpha_2, \ldots, \alpha_d]^\top$ and Q is the $d \times d$ matrix

$$Q = \begin{bmatrix} Q_{1,1} & \cdots & Q_{1,d} \\ \vdots & \ddots & \vdots \\ Q_{d,1} & \cdots & Q_{d,d} \end{bmatrix}.$$

In addition, every row of the $d \times (d + 1)$ matrix $[\alpha, Q]$ has to add up to one, i.e. we have the relation

$$\alpha_k + \sum_{l=1}^{d} Q_{k,l} = 1, \qquad k = 1, \ldots, d, \tag{2.1}$$

which is used to show the following lemma.

Lemma 2.1 *We have the relation $\alpha = (I - Q)e$, where*

$$I := \begin{bmatrix} 1 & 0 & \cdots & 0 & 0 \\ 0 & 1 & \ddots & 0 & 0 \\ \vdots & \vdots & \ddots & \ddots & \vdots \\ 0 & 0 & \cdots & 1 & 0 \\ 0 & 0 & \cdots & 0 & 1 \end{bmatrix}$$

denotes the $d \times d$ identity matrix, and

$$e := \begin{bmatrix} 1 \\ 1 \\ \vdots \\ 1 \end{bmatrix}.$$

Proof Relation (2.1) can be rewritten as

$$(I - Q)e = \begin{bmatrix} 1 - Q_{1,1} & -Q_{1,2} & \cdots & & -Q_{1,d} \\ -Q_{1,1} & 1 - Q_{1,2} & \cdots & & -Q_{1,d} \\ \vdots & \vdots & \ddots & & \vdots \\ -Q_{d,1} & & \cdots & Q_{d,d-1} & 1 - Q_{d,d} \end{bmatrix} \times \begin{bmatrix} 1 \\ 1 \\ \vdots \\ 1 \end{bmatrix}$$

$$= \begin{bmatrix} 1 - Q_{1,1} - \cdots - Q_{1,d} \\ 1 - Q_{2,1} - \cdots - Q_{2,d} \\ \vdots \\ 1 - Q_{d,1} - \cdots - Q_{d,d} \end{bmatrix}$$

$$
= \begin{bmatrix} \alpha_1 \\ \alpha_2 \\ \vdots \\ \alpha_d \end{bmatrix},
$$

which shows that $\alpha = (I - Q)e$.   □

The next proposition can be intuitively interpreted by noting that since state $\textcircled{0}$ is absorbing, the n-step behavior of the chain on the states $\{1, 2, \ldots, d\}$ is entirely determined by the matrix Q^n since when $1 \leqslant i, j \leqslant d$, as one cannot travel through state $\textcircled{0}$ when moving from $\textcircled{i}$ to $\textcircled{j}$ in any $n \geqslant 1$ number of time steps.

Proposition 2.2 *The transition matrix P of the chain $(X_n)_{n \geqslant 0}$ satisfies*

$$
P^n = \begin{bmatrix} 1 & 0 \\ (I - Q^n)e & Q^n \end{bmatrix}, \qquad n \geqslant 0. \tag{2.2}
$$

Proof We proceed by induction on $n \geqslant 0$. Clearly, the conclusion holds for $n = 0$, and also at the rank $n = 1$ since

$$
P = \begin{bmatrix} 1 & 0 \\ \alpha & Q \end{bmatrix},
$$

and $\alpha = (I - Q)e$. Next, we assume that the relation (2.2) holds at the rank $n \geqslant 0$. In this case, since

$$
\alpha + Q(I - Q^n)e = (I - Q)e + (Q - Q^{n+1})e = (I - Q^{n+1})e,
$$

we have

$$
\begin{aligned}
P^{n+1} &= P \times P^n \\
&= \begin{bmatrix} 1 & 0 \\ \alpha & Q \end{bmatrix} \times \begin{bmatrix} 1 & 0 \\ (I - Q^n)e & Q^n \end{bmatrix} \\
&= \begin{bmatrix} 1 & 0 \\ \alpha + Q(I - Q^n)e & Q^{n+1} \end{bmatrix} \\
&= \begin{bmatrix} 1 & 0 \\ (I - Q^{n+1})e & Q^{n+1} \end{bmatrix}.
\end{aligned}
$$

                                                                                              □

2.3 Hitting Time Distribution

In this section, we show that the probability distribution of the first hitting time

$$T_0 = \inf\{n \geqslant 1 \ : \ X_n = 0\}$$

of state $\textcircled{0}$ after starting from state $i \geqslant 1$ can be computed using the vector α and the matrix Q.

Proposition 2.3 *For all $i = 1, 2, \ldots, d$ we have*

$$\mathbb{P}\big(T_0 = n \mid X_0 = i\big) = [Q^{n-1}\alpha]_i, \qquad n \geqslant 1. \tag{2.3}$$

Proof Since the state $\textcircled{0}$ is absorbing, we can partition the event $\{T_0 = n\}$ as

$$\{T_0 = n\} = \bigcup_{k=1}^{d} \{X_n = 0 \text{ and } X_{n-1} = k\},$$

and note that, since $[P^{n-1}]_{i,k} = [Q^{n-1}]_{i,k}$ from Proposition 2.2 and $\alpha_k = P_{k,0}$, $k = 1, 2, \ldots, d$, we have

$$\mathbb{P}\big(T_0 = n \mid X_0 = i\big) = \mathbb{P}\left(\bigcup_{k=1}^{d} \{T_0 = n, \ X_{n-1} = k\} \ \bigg| \ X_0 = i\right)$$

$$= \sum_{k=1}^{d} \mathbb{P}(T_0 = n, \ X_{n-1} = k \mid X_0 = i)$$

$$= \sum_{k=1}^{d} \mathbb{P}(X_n = 0, \ X_{n-1} = k \mid X_0 = i)$$

$$= \sum_{k=1}^{d} \mathbb{P}(X_n = 0 \mid X_{n-1} = k)\mathbb{P}(X_{n-1} = k \mid X_0 = i)$$

$$= \sum_{k=1}^{d} [P^{n-1}]_{i,k} P_{k,0}$$

$$= \sum_{k=1}^{d} \alpha_k [Q^{n-1}]_{i,k}$$

$$= [Q^{n-1}\alpha]_i, \qquad n \geqslant 1.$$

$\square$

From now on, we assume that the initial distribution of X_0 on $\{1, 2, \ldots, d\}$ is given by the d-dimensional vector

$$\beta = \begin{bmatrix} \beta_1 \\ \beta_2 \\ \vdots \\ \beta_d \end{bmatrix},$$

i.e.

$$\beta_i = \mathbb{P}(X_0 = i), \qquad i = 1, 2, \ldots, d,$$

with $\mathbb{P}(X_0 = 0) = 0$.

Proposition 2.4 *The probability distribution of T_0 is given by*

$$\mathbb{P}(T_0 = n) = \beta^\top Q^{n-1} \alpha, \qquad n \geqslant 1.$$

Proof By (2.3), we have

$$\mathbb{P}(T_0 = n) = \sum_{i=1}^{d} \mathbb{P}(T_0 = n \mid X_0 = i) \mathbb{P}(X_0 = i)$$

$$= \sum_{i=1}^{d} \beta_i [Q^{n-1} \alpha]_i$$

$$= \sum_{i=1}^{d} \beta_i \sum_{k=1}^{d} \alpha_k Q_{i,k}^{n-1}$$

$$= \beta^\top Q^{n-1} \alpha, \qquad n \geqslant 1.$$

$\square$

Since the states $\{1, 2, \ldots, d\}$ are transient, Corollary 1.15 shows that the matrix inverse $(I - sQ)^{-1}$ exists and is given by the series

$$(I - sQ)^{-1} = \sum_{k \geqslant 0} s^k Q^k, \qquad s \in (-1, 1]. \tag{2.4}$$

We note that T_0 is finite with probability one, since

$$
\begin{aligned}
\mathbb{P}(T_0 < \infty) &= \sum_{n=0}^{\infty} \mathbb{P}(T_0 = n) \\
&= \sum_{n=1}^{\infty} \beta^{\top} Q^{n-1} \alpha \\
&= \beta^{\top} \sum_{n=0}^{\infty} Q^n \alpha \\
&= \beta^{\top} (I - Q)^{-1} \alpha \\
&= \beta^{\top} (I - Q)^{-1} (I - Q) e \\
&= \sum_{i=1}^{d} \beta_i \\
&= \sum_{i=1}^{d} \mathbb{P}(X_0 = i) \\
&= 1.
\end{aligned}
$$

Corollary 2.5 *The cumulative distribution function $P(T_0 \leqslant n)$ of T_0 is given in terms of the vectors β, e, and the matrix Q^n as*

$$
\mathbb{P}(T_0 \leqslant n) = 1 - \beta^{\top} Q^n e, \qquad n \geqslant 0. \tag{2.5}
$$

Proof Using the relation $\alpha = (I - Q)e$, we have

$$
\begin{aligned}
\mathbb{P}(T_0 \leqslant n) &= \sum_{k=1}^{n} \mathbb{P}(T_0 = k) \\
&= \sum_{k=1}^{n} \beta^{\top} Q^{k-1} \alpha \\
&= \beta^{\top} (I - Q^n)(I - Q)^{-1} \alpha \\
&= \beta^{\top} (I - Q^n) e \\
&= 1 - \beta^{\top} Q^n e, \qquad n \geqslant 1.
\end{aligned}
$$

$\square$

Alternatively, also using the relation $\alpha = (I - Q)e$ and a telescopic sum, Relation (2.5) can be recovered as

$$\mathbb{P}(T_0 \leqslant n) = \sum_{k=1}^{n} \mathbb{P}(T_0 = k)$$

$$= \sum_{k=1}^{n} \beta^\top Q^{k-1} \alpha$$

$$= \sum_{k=1}^{n} \beta^\top Q^{k-1} (I - Q)e$$

$$= \sum_{k=0}^{n-1} \beta^\top Q^k e - \sum_{k=1}^{n} \beta^\top Q^k e$$

$$= \beta^\top e - \beta^\top Q^n e$$

$$= 1 - \beta^\top Q^n e, \qquad n \geqslant 1.$$

We can also rewrite $\mathbb{P}(T_0 \leqslant n)$ as the probability of not being in any state $i = 1, 2, \ldots, d$ at time n, as

$$\mathbb{P}(T_0 \leqslant n) = 1 - \sum_{k=1}^{d} \mathbb{P}(X_n = k)$$

$$= 1 - \sum_{k=1}^{d} \sum_{i=1}^{d} \beta_i \mathbb{P}(X_n = k \mid X_0 = i)$$

$$= 1 - \sum_{k=1}^{d} \sum_{i=1}^{d} \beta_i Q_{i,k}^n$$

$$= 1 - \beta^\top Q^n e, \qquad n \geqslant 0.$$

Alternatively, we could also write

$$\mathbb{P}(T_0 \leqslant n) = \mathbb{P}(X_n = 0)$$

$$= \sum_{i=1}^{d} \beta_i \mathbb{P}(X_n = 0 \mid X_0 = i)$$

$$= \sum_{i=1}^{d} \beta_i [P^n]_{i,0}$$

$$= \sum_{i=1}^{d} \beta_i [(I - Q^n)e]_i$$

$$= \sum_{i=1}^{d} \beta_i - \sum_{i=1}^{d} \beta_i [Q^n e]_i$$

$$= 1 - \beta^\top Q^n e, \qquad n \geqslant 0.$$

We refer to the Appendix for the definition of the Probability Generating Function (PGF) of a discrete random variable.

Proposition 2.6 *The probability generating function*

$$G_{T_0}(s) := \mathbb{E}\big[s^{T_0} \mathbb{1}_{\{T_0 < \infty\}}\big] = \sum_{k \geqslant 0} s^k \mathbb{P}(T_0 = k)$$

of T_0 is given by

$$G_{T_0}(s) = s\beta^\top (I - sQ)^{-1}(I - Q)e. \tag{2.6}$$

Proof By (2.7) we have $\mathbb{P}(T_0 < \infty) = 1$, hence

$$G_{T_0}(s) = \sum_{k \geqslant 0} s^k \mathbb{P}(T_0 = k)$$

$$= \mathbb{P}(X_0 = 0) + \sum_{k \geqslant 1} s^k \beta^\top Q^{k-1} \alpha$$

$$= s \sum_{k \geqslant 0} s^k \beta^\top Q^k \alpha$$

$$= s\beta^\top \sum_{k \geqslant 0} s^k Q^k \alpha$$

$$= s\beta^\top (I - sQ)^{-1} \alpha$$

$$= s\beta^\top (I - sQ)^{-1}(I - Q)e,$$

where we applied Lemma 2.1 and (2.4). □

We note that

$$\mathbb{P}(T_0 < \infty) = G_{T_0}(1) = \beta^\top (I - Q)^{-1}(I - Q)e = \beta^\top e = 1, \tag{2.7}$$

which shows that state $\textcircled{0}$ is reached in finite time with probability one.

2.4   Mean Hitting Times

Using the probability generating function $s \mapsto G_{T_0}(s)$, we compute the first and second moments $E[T_0]$ and $E[T_0^2]$ of T_0. By differentiating (2.6) with respect to s we have

$$G'_{T_0}(s) = \beta^\top (I - sQ)^{-1}\alpha + s\beta^\top Q(I - sQ)^{-2}\alpha,$$

hence[2]

$$
\begin{aligned}
\mathbb{E}[T_0] &= G'_{T_0}(1^-) \\
&= \beta^\top (I - Q)^{-1}\alpha + \beta^\top Q(I - Q)^{-2}\alpha \\
&= \beta^\top (I - Q)(I - Q)^{-2}\alpha + \beta^\top Q(I - Q)^{-2}\alpha \\
&= \beta^\top (I - Q)^{-2}\alpha \\
&= \beta^\top (I - Q)^{-1}e.
\end{aligned}
$$

By differentiating (2.6) further, we also have

$$G''_{T_0}(s) = \beta^\top Q(I - sQ)^{-2}\alpha + \beta^\top Q(I - sQ)^{-2}\alpha + 2s\beta^\top Q^2(I - sQ)^{-3}\alpha,$$

hence

$$
\begin{aligned}
\mathbb{E}[T_0(T_0 - 1)] &= G''_{T_0}(1^-) \\
&= 2\beta^\top Q(I - Q)^{-2}\alpha + 2\beta^\top Q^2(I - Q)^{-3}\alpha \\
&= 2\beta^\top Q(I - Q)^{-3}\alpha, \\
&= 2\beta^\top Q(I - Q)^{-2}e,
\end{aligned}
$$

and

$$
\begin{aligned}
\mathbb{E}[T_0^2] &= \mathbb{E}[T_0(T_0 - 1)] + \mathbb{E}[T_0] \\
&= 2\beta^\top Q(I - Q)^{-2}e + \beta^\top (I - Q)^{-1}e \\
&= 2\beta^\top Q(I - Q)^{-2}e + \beta^\top (I - Q)(I - Q)^{-2}e \\
&= \beta^\top (I + Q)(I - Q)^{-2}e.
\end{aligned}
$$

---

[2] Here, $G'_X(1^-)$ denotes the derivative on the left at the point $s = 1$.

More generally, by (A.4) we could also compute the *factorial moment*

$$\mathbb{E}[T_0(T_0 - 1) \cdots (T_0 - k + 1)] = G_{T_0}^{(k)}(1^-) = k! \beta^\top Q^{k-1}(I - Q)^{-k} e,$$

for all $k \geqslant 1$.

Notes

See e.g. Latouche and Ramaswami (1999) for further reading.

Exercises

Exercise 2.1 (Vinay and Kok (2019)) . The double-heralding protocol for entanglement generation in quantum cryptography involves two rounds of photon transfer, the failure of either of which will cause the process to be restarted.

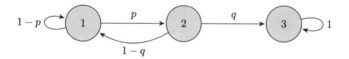

The protocol is modeled using a Markov chain $(X_n)_{n \geqslant 0}$ described by the above graph, in which $p \in (0, 1)$ is the probability of passing the first round, and $q \in (0, 1)$ is the probability of passing the second round, conditional on passing the first. Let

$$T_3 = \inf\{n \geqslant 0 : X_n = 3\}$$

denote the first hitting time of state ③ by the chain $(X_n)_{n \geqslant 0}$.

(a) Using first step analysis, find the mean time to completion of double-heralding after starting from state ⓘ, $i = 1, 2$.
(b) Find the Probability Generating Function (PGF)

$$G_i(s) = \mathbb{E}[s^{T_3} \mid X_0 = i], \qquad -1 \leqslant s \leqslant 1,$$

of T_3 after starting from state ⓘ, $i = 1, 2$.
Hint. Start by deriving a system of equations satisfied by $G_i(s)$ using first step analysis.
(c) Find the probability distribution $\mathbb{P}(T_3 = k \mid X_0 = 1)$ of the completion time after starting from state ①.

Hint. Use the power series expansion

$$\frac{\sqrt{(1-p)^2 + 4(1-q)p}}{1 - (1-p)s - p(1-q)s^2} = \sum_{n=0}^{\infty} \frac{s^n}{z_+^{n+1}} - \sum_{n=0}^{\infty} \frac{s^n}{z_-^{n+1}},$$

where

$$z_\pm := \frac{p - 1 \pm \sqrt{(1-p)^2 + 4(1-q)p}}{2(1-q)p}.$$

(d) Using $G_1(s)$, recover the mean time to completion of double-heralding after starting from state $\textcircled{1}$, as found in Question (a).

Chapter 3
Synchronizing Automata

Synchronizing automata are connected to algebra and combinatorics, and they have applications in many areas including robotics, coding theory, network security management, chip design, industrial automation, biocomputing, etc. In this chapter, we consider synchronizing automata in the framework of Markovian text generation, with examples of application to pattern recognition in randomly generated sequences.

3.1 Pattern Recognition

Given an alphabet made of a finite set Σ of letters, we denote by Σ^* the set of all (finite) words over Σ, i.e. Σ^* is made of all finite sequences of symbols in Σ.

Definition 3.1 A *language* $\mathcal{L}$ over a set Σ of letters is a collection of (finite) words in Σ^*. The notation $\Sigma^* xxxxx \Sigma^*$ denotes the concatenations of a word in Σ^* followed by a certain word $xxxxx$, followed by another word in Σ^*.

Markovian Text Generation

We would like to determine the mean time until a certain character string appears in a random sequence generated by a Markov chain.

© The Author(s), under exclusive license to Springer Nature Switzerland AG 2024
N. Privault, *Discrete Stochastic Processes*, Springer Undergraduate
Mathematics Series, https://doi.org/10.1007/978-3-031-65820-4_3

First-Order Word Analysis

The following ℝ code is estimating a transition matrix P for the first order analysis of a text of 10000 characters.

```
text = readChar("text_file.txt",nchars=10000)
x <- unlist (strsplit (gsub ("[^a-z]", "-", tolower (text)), ""))
P <- matrix(nrow = 27, ncol = 27, 0,dimnames = list(c("-", letters),c("-", letters)))
for (t in 1:(length(x) - 1)) P[x[t], x[t + 1]] <- P[x[t], x[t + 1]] + 1
for (i in 1:27) P[i, ] <- P[i, ] / sum(P[i, ])
P[1:5,1:5]
```

The transition matrix P is estimated by counting the proportion of transitions from any given state $\widehat{i}$ to another state $\widehat{j}$, $i, j \in S$, using

$$\widehat{P}_{i,j}(m) := \frac{1}{R_i(m)} \sum_{t=1}^{m-1} \mathbb{1}_{\{X_t=i,\ X_{t+1}=j\}},$$

where

$$R_i(m) := \sum_{n=1}^{m-1} \mathbb{1}_{\{X_n=j\}}, \qquad m \geqslant 2,$$

denotes the number of returns to state $\widehat{j}$ by the chain $(X_n)_{n \geqslant 0}$ up to time $m - 1$. Next is a sample transition matrix obtained from this analysis.

$$
\begin{array}{c}
 \\ - \\ a \\ b \\ c \\ d
\end{array}
\begin{array}{ccccc}
\ - & a & b & c & d \\
\left[\begin{array}{ccccc}
0.43959353 & 0.08882198 & 0.004140008 & 0.01204366 & 0.01656003 \cdots \\
0.11014493 & 0.00000000 & 0.031884058 & 0.00000000 & 0.14057971 \cdots \\
0.08461538 & 0.00000000 & 0.000000000 & 0.00000000 & 0.00000000 \cdots \\
0.04803493 & 0.04803493 & 0.000000000 & 0.00000000 & 0.00000000 \cdots \\
0.74644550 & 0.02606635 & 0.000000000 & 0.00000000 & 0.02369668 \cdots \\
\vdots & \vdots & \vdots & \vdots & \vdots & \ddots
\end{array}\right]
\end{array}
$$

```
install.packages("devtools"); library(devtools) # Install RTools as well
devtools::install_github('spedygiorgio/markovchain')
install.packages("igraph"); library(igraph);library(markovchain)
MC <-new("markovchain", transitionMatrix=P,states=c("-", letters))
graph <- as(MC, "igraph")
plot(graph,edge.label.cex=1,edge.label=sprintf("%1.2f",
     E(graph)$prob),edge.color='black', vertex.color='dodgerblue',vertex.label.cex=1)
cat(markovchainSequence(n = 100, markovchain = MC, t0 = "a", include.t0 = TRUE),"\n" )
```

Second-Order Word Analysis

For simplicity, we consider a two-state Markov chain $(X_n)_{n \geqslant 0}$ taking values in the two-letter alphabet $S = \{a, b\}$ and transition matrix

$$P = \begin{array}{c} \\ a \\ b \end{array} \begin{array}{cc} a & b \\ \left[\begin{array}{cc} 1 - q & q \\ p & 1 - p \end{array} \right], \end{array}$$

where $p, q \in (0, 1)$.

Next, we define a new stochastic process $(Z_n)_{n \geqslant 1}$ by $Z_n = (X_{n-1}, X_n), n \geqslant 1$, which models words of length (or order) 2. The state space of $(Z_n)_{n \geqslant 1}$ is made of the set of words

$$\{(a, a), (a, b), (b, a), (b, b)\}$$

which corresponds to two-step text generation. Based on $Z_n = (X_{n-1}, X_n)$, the distribution of $Z_{n+1} = (X_n, X_{n+1})$ at time $n + 1$ is fully determined from the data of X_n and the transition matrix of $(X_n)_{n \geqslant 0}$ hence $(Z_n)_{n \geqslant 1}$ is a $\{aa, ab, ba, bb\}$-valued Markov chain, whose transition matrix is given by

$$\begin{array}{c} \\ aa \\ ab \\ ba \\ bb \end{array} \begin{array}{cccc} aa & ab & ba & bb \\ \left[\begin{array}{cccc} 1 - q & q & 0 & 0 \\ 0 & 0 & p & 1 - p \\ 1 - q & q & 0 & 0 \\ 0 & 0 & p & 1 - p \end{array} \right]. \end{array} \qquad (3.1)$$

- Starting from $Z_n = (a, a)$, if the next letter is $X_{n+1} = a$ (with probability p) then we obtain

$$(X_{n-1}, X_n, X_{n+1}) = (a, a, a).$$

The next 2-letter state Z_{n+1} is now based on the last two letters of (a, a, a), i.e. $Z_{n+1} = (a, a)$. In this case, the 2-letter state switches from $Z_n = (a, a)$ to $Z_{n+1} = (a, a)$ with probability p.

- Starting from $Z_n = (a, a)$, if the next letter is $X_{n+1} = b$ (with probability q) then we obtain

$$(X_{n-1}, X_n, X_{n+1}) = (a, a, b).$$

The current 2-letter state is now based on the last two letters of (a, a, b), i.e. $Z_{n+1} = (a, b)$. In this case, the 2-letter state switches from $Z_n = (a, a)$ to $Z_{n+1} = (a, b)$ with probability q.

- Starting from $Z_n = (b, a)$, if the next letter is $X_{n+1} = a$ (with probability p) then we obtain

$$(X_{n-1}, X_n, X_{n+1}) = (b, a, a).$$

The current 2-letter state is now based on the last two letters of (b, a, a), i.e. $Z_{n+1} = (a, a)$. In this case, the 2-letter state switches from $Z_n = (b, a)$ to $Z_{n+1} = (a, a)$ with probability p.

- Starting from $Z_n = (b, a)$, if the next letter is $X_{n+1} = b$ (with probability q) then we obtain

$$(X_{n-1}, X_n, X_{n+1}) = (b, a, b).$$

The current 2-letter state is now based on the last two letters of (b, a, b), i.e. $Z_{n+1} = (a, b)$. In this case, the 2-letter state switches from $Z_n = (b, a)$ to $Z_{n+1} = (a, b)$ with probability q.

- On the other hand, starting from $Z_n = (x, a)$, resp. $Z_n = (x, b)$, we cannot switch to any state of the form $Z_{n+1} = (b, y)$, resp. $Z_n = (a, y)$, by construction of $Z_n := (X_{n-1}, X_n)$, $x, y \in \{a, b\}$.

A similar reasoning can be applied to other entries in the transition matrix (3.1). The text_generation_page_64.ipynb Python code uploaded as supplementary material implements the estimation of transition matrix for any order, and generates the corresponding chain samples.

Independent Samples

In what follows, we assume that $p + q = 1$ with $0 < p < 1$, in which case the transition matrix P becomes

$$P = \begin{array}{c} a \\ b \end{array} \begin{array}{cc} a & b \\ \left[\begin{array}{cc} p & q \\ p & q \end{array} \right], \end{array}$$

and the sequence $(X_n)_{n \geqslant 0}$ is made of identically distributed Bernoulli random variables taking values in $\{a, b\}$, such that

$$\mathbb{P}(X_n = a) = p \quad \text{and} \quad \mathbb{P}(X_n = b) = q = 1 - p, \quad n \geqslant 0. \tag{3.2}$$

In addition, the sequence X_n is independent of $X_{n+1}, n \geqslant 0$, as

$$\mathbb{P}(X_{n+1} = x) = \sum_{z \in \{a,b\}} \mathbb{P}(X_{n+1} = x \mid X_n = z)\mathbb{P}(X_n = z)$$

$$= \sum_{z \in \{a,b\}} \mathbb{P}(X_{n+1} = x \mid X_n = y)\mathbb{P}(X_n = z)$$

$$= \mathbb{P}(X_{n+1} = x \mid X_n = y) \sum_{z \in \{a,b\}} \mathbb{P}(X_n = z)$$

$$= \mathbb{P}(X_{n+1} = x \mid X_n = y), \qquad y \in \{a, b\},$$

which shows that

$$\mathbb{P}(X_{n+1} = x \text{ and } X_n = y) = \mathbb{P}(X_{n+1} = x \mid X_n = y)\mathbb{P}(X_n = y)$$
$$= \mathbb{P}(X_{n+1} = x)\mathbb{P}(X_n = y), \quad x, y \in \{a, b\}.$$

We note that $(Z_n)_{n \geqslant 1} = ((X_{n-1}, X_n))_{n \geqslant 1}$ is a Markov chain with four possible states denoted $\{aa, ab, ba, bb\}$, and write down its 4×4 transition matrix. Precisely, the transition matrix of $(Z_n)_{n \geqslant 1}$ is given by

$$\begin{array}{c} & \begin{array}{cccc} aa & ab & ba & bb \end{array} \\ \begin{array}{c} aa \\ ab \\ ba \\ bb \end{array} & \left[\begin{array}{cccc} p & q & 0 & 0 \\ 0 & 0 & p & q \\ p & q & 0 & 0 \\ 0 & 0 & p & q \end{array} \right] \end{array}.$$

Average Recognition Times

Let now

$$\tau_{ab} := \inf\{n \geqslant 1 \ : \ Z_n = (a, b)\}$$

denote the first time of appearance of the pattern "*ab*" in the sequence $(X_0, X_1, X_2, \ldots)$. The mean time it takes until we encounter the pattern "*ab*" after starting from $X_0 = a$ can be computed as a consequence of Proposition 1.3, as

$$\mathbb{E}[\tau_{ab} \mid X_0 = a] = \frac{1}{q} = 1 + \frac{p}{q}. \tag{3.3}$$

This mean time can be recovered by pathwise analysis using the mean $1/q$ of the geometric distribution on $\{1, 2, 3, \ldots\}$ with parameter $q \in (0, 1]$, as

$$
\begin{aligned}
\mathbb{E}[\tau_{ab} \mid X_0 = a] &= \sum_{n=1}^{\infty} n p^{n-1} q \\
&= \sum_{n=0}^{\infty} (n+1) p^n q \\
&= q \sum_{n=0}^{\infty} p^n + p \sum_{n=0}^{\infty} n p^{n-1} q \\
&= 1 + \frac{pq}{(1-p)^2} \\
&= 1 + \frac{p}{q} \\
&= \frac{1}{q}.
\end{aligned}
\tag{3.4}
$$

Given the initial value of Z_1 we can also compute the probability distribution

$$
\mathbb{P}(\tau_{ab} = n \mid Z_1 = (a, a)) = \mathbb{P}(\tau_{ab} = n \mid Z_1 = (b, a)) = q p^{n-2}, \qquad n \geqslant 2,
\tag{3.5}
$$

of the hitting time τ_{ab} after starting from either (a, a) or (b, a), according to the following examples:

$$
\underset{\substack{\uparrow \ \uparrow \ \uparrow \ \uparrow \ \uparrow \\ 0 \ \ 1 \ \ 2 \ \ 3 \ \ 4}}{(a, a, a, a, a, \ldots,} \ \underset{\substack{\uparrow \qquad \uparrow \\ n-1 \quad n}}{a \ , b),} \qquad \underset{\substack{\uparrow \ \uparrow \ \uparrow \ \uparrow \ \uparrow \\ 0 \ \ 1 \ \ 2 \ \ 3 \ \ 4}}{(b, a, a, a, a, \ldots,} \ \underset{\substack{\uparrow \qquad \uparrow \\ n-1 \quad n}}{a \ , b).}
$$

Probability Generating Functions

In the remainder of this section we we consider an alternative approach using the probability generating functions

$$
G_{aa}(s) := \mathbb{E}\left[s^{\tau_{ab}} \mathbb{1}_{\{\tau_{ab} < \infty\}} \mid Z_1 = (a, a) \right], \qquad -1 \leqslant s \leqslant 1,
$$

and

$$
G_{ba}(s) := \mathbb{E}\left[s^{\tau_{ab}} \mathbb{1}_{\{\tau_{ab} < \infty\}} \mid Z_1 = (b, a) \right], \qquad -1 \leqslant s \leqslant 1,
$$

which satisfy

$$G_{aa}(s) = G_{ba}(s), \qquad -1 < s < 1.$$

We also note that the probability generating function

$$G_{ab}(s) := \mathbb{E}\big[s^{\tau_{ab}} \mathbb{1}_{\{\tau_{ab}<\infty\}} \mid Z_1 = (a, b)\big], \qquad -1 \leqslant s \leqslant 1.$$

satisfies

$$G_{ab}(s) = \mathbb{E}[s \mid Z_1 = (a, b)] = s, \qquad -1 \leqslant s \leqslant 1,$$

since given that $Z_1 = (a, b)$ we have $\tau_{ab} = 1$ with probability one.

Proposition 3.2 *The Probability Generating Function (PGF) of the hitting time τ_{ab} satisfies*

$$G_{aa}(s) = G_{ba}(s) = \frac{qs^2}{1 - ps}, \qquad -1 \leqslant s \leqslant 1. \tag{3.6}$$

Proof Using first step analysis, we have

$$\begin{aligned}
G_{aa}(s) &= \mathbb{E}\big[s^{\tau_{ab}} \mathbb{1}_{\{\tau_{ab}<\infty\}} \mid Z_1 = (a, a)\big] \\
&= p\mathbb{E}\big[s^{\tau_{ab}} \mathbb{1}_{\{\tau_{ab}<\infty\}} \mid Z_2 = (a, a)\big] + q\mathbb{E}\big[s^{\tau_{ab}} \mathbb{1}_{\{\tau_{ab}<\infty\}} \mid Z_2 = (a, b)\big] \\
&= p\mathbb{E}\big[s^{1+\tau_{ab}} \mathbb{1}_{\{\tau_{ab}<\infty\}} \mid Z_1 = (a, a)\big] + q\mathbb{E}\big[s^{1+\tau_{ab}} \mathbb{1}_{\{\tau_{ab}<\infty\}} \mid Z_1 = (a, b)\big]
\end{aligned}$$

and

$$\begin{aligned}
G_{ba}(s) &= \mathbb{E}\big[s^{\tau_{ab}} \mathbb{1}_{\{\tau_{ab}<\infty\}} \mid Z_1 = (b, a)\big] \\
&= p\mathbb{E}\big[s^{\tau_{ab}} \mathbb{1}_{\{\tau_{ab}<\infty\}} \mid Z_2 = (a, a)\big] + q\mathbb{E}\big[s^{\tau_{ab}} \mathbb{1}_{\{\tau_{ab}<\infty\}} \mid Z_2 = (a, b)\big] \\
&= p\mathbb{E}\big[s^{1+\tau_{ab}} \mathbb{1}_{\{\tau_{ab}<\infty\}} \mid Z_1 = (a, a)\big] + q\mathbb{E}\big[s^{1+\tau_{ab}} \mathbb{1}_{\{\tau_{ab}<\infty\}} \mid Z_1 = (a, b)\big],
\end{aligned}$$

which yields the system of equations

$$\begin{cases} G_{aa}(s) = psG_{aa}(s) + qsG_{ab}(s) \\ \\ G_{ba}(s) = psG_{aa}(s) + qsG_{ab}(s), \end{cases} \tag{3.7}$$

where $G_{ab}(s) = \mathbb{E}[s \mid Z_1 = (a, b)] = s, -1 \leqslant s \leqslant 1$. Therefore, we have

$$\begin{cases} G_{aa}(s) = psG_{aa}(s) + qs^2, \\ \\ G_{ba}(s) = psG_{aa}(s) + qs^2, \end{cases}$$

from which we compute $G_{aa}(s)$ and $G_{ba}(s)$ as

$$G_{aa}(s) = G_{ba}(s) = \frac{pqs^3}{1-ps} + qs^2 = \frac{qs^2}{1-ps}, \qquad -1 \leqslant s \leqslant 1.$$

$$\square$$

From Proposition 3.2, we note that

$$\mathbb{P}(\tau_{ab} < \infty \mid Z_1 = (a,a)) = \mathbb{P}(\tau_{ab} < \infty \mid Z_1 = (b,a))$$
$$= G_{aa}(1) = G_{ba}(1)$$
$$= \frac{q}{1-p}$$
$$= 1.$$

In addition, by expanding the PGF in (3.6) into the series

$$G_{aa}(s) = \frac{qs^2}{1-ps}$$
$$= qs^2 \sum_{k \geqslant 0} p^k s^k$$
$$= q \sum_{k \geqslant 2} p^{k-2} s^k$$
$$= \sum_{k \geqslant 0} s^k \mathbb{P}(\tau_{ab} = k \mid Z_1 = (a,a))$$

recovers the probability distribution (3.5). The probability generating functions can now be used to compute the mean times

$$\mathbb{E}[\tau_{ab} \mid Z_1 = (a,a)] \quad \text{and} \quad \mathbb{E}[\tau_{ab} \mid Z_1 = (b,a)],$$

as

$$\mathbb{E}[\tau_{ab} \mid Z_1 = (a,a)] = \mathbb{E}[\tau_{ab} \mid Z_1 = (b,a)]$$
$$= G'_{ba}(1^-) = G'_{aa}(1^-)$$
$$= \frac{2q}{1-p} + \frac{pq}{(1-p)^2}$$
$$= 2 + \frac{p}{q} = 1 + \frac{1}{q},$$

which is consistent with (3.3)–(3.4) as one time step is needed to switch from $X_0 = a$ to $X_1 = a$ when $Z_1 = (a, a)$. The next proposition recovers (3.3) using probability generating functions.

Proposition 3.3 *The average time* $\mathbb{E}[\tau_{ab} \mid X_0 = a]$ *it takes until we encounter the pattern "ab" in the sequence* $(X_0, X_1, X_2, \ldots)$ *started with* $X_0 = a$ *is* $1 + p/q$.

Proof The average time it takes until we encounter the pattern "*ab*" in the sequence $(X_0, X_1, X_2, \ldots)$ started with $X_0 = a$ is given by

$$\mathbb{E}[\tau_{ab} \mid X_0 = a] = p\mathbb{E}[\tau_{ab} \mid Z_1 = (a, a)] + q\mathbb{E}[\tau_{ab} \mid Z_1 = (a, b)]$$
$$= p\left(2 + \frac{p}{q}\right) + q$$
$$= 1 + \frac{p}{q}.$$

$\square$

The next section illustrates the use of probability generating functions in more complex situations.

3.2 Winning Streaks

Consider a sequence $(X_n)_{n \geqslant 1}$ of independent Bernoulli random variables with the distribution

$$\mathbb{P}(X_n = a) = p, \qquad \mathbb{P}(X_n = b) = q, \qquad n \geqslant 1,$$

with $q := 1 - p$. For some $m \geqslant 1$, let $T^{(m)}$ denote the time of the first appearance of m consecutive a's in the sequence $(X_n)_{n \geqslant 1}$. For example, taking $m := 4$, the sequence

$$
\overset{\displaystyle \text{4 times}}{(b, a, a, b, \overbrace{a, a, a, a}, b, a, a, b, \ldots)}
$$
$$\underset{1 \quad 2 \quad 3 \quad 4 \quad 5 \quad 6 \quad 7 \quad 8}{\uparrow \quad \uparrow \quad \uparrow \quad \uparrow \quad \uparrow \quad \uparrow \quad \uparrow \quad \uparrow}$$

yields $T^{(4)} = 8$.

Probability Distribution of $T^{(m)}$

We note that

(a) We have $\mathbb{P}\big(T^{(m)} < m\big) = 0$ since it takes at least m letters to form an m-winning streak.
(b) We have $\mathbb{P}\big(T^{(m)} = m\big) = p^m$ since observing an m-winning streak at time m requires to generate exactly exactly m times "a".
(c) We have $\mathbb{P}\big(T^{(m)} = m+1\big) = qp^m$ because observing the first m-winning streak at time $m + 1$ exactly requires to generate the sequence

$$
(b, \overbrace{a, a, a, \ldots, a}^{m \text{ times}}).
$$
$$
\underset{4 \quad 5 \quad 6 \quad 7 \qquad\quad m+1}{\uparrow\ \uparrow\ \uparrow\ \uparrow \qquad\quad \uparrow}
$$

(d) We have

$$
\mathbb{P}\big(T^{(m)} = m+2\big) = q^2 p^m + pqp^m = qp^m
$$

because observing the first m-winning streak at time $m + 2$ can be achieved *via* exactly two sequences

$$
(b, b, \overbrace{a, a, a, \ldots, a}^{m \text{ times}}) \qquad \text{and} \qquad (a, b, \overbrace{a, a, a, \ldots, a}^{m \text{ times}}).
$$

(e) More generally, for $n = 1, 2, \ldots, m$ we find

$$
\mathbb{P}\big(T^{(m)} = n+m\big) = qp^m = qp^m \sum_{k=0}^{n-1} \binom{n-1}{k} q^k p^{n-1-k},
$$

because when $n \leqslant m$ any sequence of the form

$$
(\overbrace{x_1, x_2, \ldots, x_{n-1}}^{n-1 \text{ times}}, b, \overbrace{a, \ a, \ a, \ldots, \ a}^{m \text{ times}}, \ldots)
$$

$x_1, x_2, \ldots, x_{n-1} \in \{a, b\}$, will generate an m-winning streak at time $n + m$.

In the general case, computing $\mathbb{P}\big(T^{(m)} = k\big)$ for $k \geqslant 2m + 1$ is more difficult, see Exercise 3.6. In the Proposition 3.4 we compute the probability generating function

$$
G_{T^{(m)}}(s) := \mathbb{E}\big[s^{T^{(m)}} \mathbb{1}_{\{T^{(m)} < \infty\}}\big], \qquad -1 \leqslant s \leqslant 1.
$$

Proposition 3.4 *The probability generating function $G_{T^{(m)}}(s)$ satisfies*

$$G_{T^{(m)}}(s) = \frac{p^m s^m (1 - ps)}{1 - s + qp^m s^{m+1}}, \qquad -1 \leqslant s \leqslant 1, \quad m \geqslant 1. \tag{3.8}$$

Proof We apply a "k-step analysis" argument to all possible starting patterns of the form

$$\underbrace{(a, a, \ldots, a}_{k \text{ times}}, b, \ldots)$$
$$\;\uparrow\;\;\uparrow\qquad\;\;\uparrow\;\;\uparrow$$
$$\;1\;\;2\qquad\;\;k\;\;k+1$$

where $k = 0, 1, \ldots, m$, i.e.

$$\begin{array}{ll} b & k = 0 \\ ab & k = 1 \\ aab & k = 2 \\ \;\;\vdots & \\ a \cdots ab & k = m - 1 \\ \underbrace{a \cdots aa}_{m \text{ times}} & k = m, \end{array}$$

and we compute their respective probabilities. The idea is to start by flipping a coin and to observe the number k of consecutive "a" until we get the first "b".

1. If $k = m$ then the game ends, and this happens with probability $\mathbb{P}(T^{(m)} = m) = p^m$.
2. If $k < m$, the sequence of "a" is broken and we need to start again at time $k + 1$. This happens with probability $p^k q$ and we need to factor in the power s^{k+1} where $k + 1$ is the number of time steps until we reach the first "b", and restart the counter $T^{(m)}$.

In other words, we have

$$G_{T^{(m)}}(s) = s^m \mathbb{P}(T^{(m)} = m) + \sum_{k=0}^{m-1} qp^k \mathbb{E}[s^{k+1+T^{(m)}}]$$

$$= p^m s^m + \sum_{k=0}^{m-1} p^k q s^{k+1} \mathbb{E}[s^{T^{(m)}}]$$

$$= p^m s^m + qs G_{T^{(m)}}(s) \sum_{k=0}^{m-1} (ps)^k \tag{3.9}$$

$$= p^m s^m + qs G_{T^{(m)}}(s) \frac{1 - (ps)^m}{1 - ps}, \qquad -1 \leqslant s \leqslant 1,$$

which yields (3.8), where we used the relation

$$\sum_{k=0}^{m-1} x^k = \frac{1 - x^m}{1 - x}, \qquad x \in (-1, 1).$$

□

We note that

$$\mathbb{P}\big(T^{(m)} < \infty\big) = G_{T^{(m)}}(1) = (1 - p)\frac{p^m}{qp^m} = 1, \tag{3.10}$$

hence the time $T^{(m)}$ until the first m-winning streak is finite with probability one.

Next, from the probability generating function $G_{T^{(m)}}(s)$, we compute the mean time $\mathbb{E}[T^{(m)}]$ until we encounter an m-winning streak, for all $m \geqslant 1$. See also Exercises 3.1 and 3.2 for alternative methods.

Proposition 3.5 *We have*

$$\mathbb{E}\big[T^{(m)}\big] = G'_{T^{(m)}}(1) = \frac{1 - p^m}{(1 - p)p^m} = \frac{(1/p)^m - 1}{1 - p} = \sum_{k=1}^{m} \frac{1}{p^k}, \quad m \geqslant 1.$$
$$\tag{3.11}$$

Proof Instead of differentiating (3.8) it can be simpler to differentiate (3.9) with respect to s, which yields

$$G'_{T^{(m)}}(s) = mp^m s^{m-1} + qG'_{T^{(m)}}(s) \sum_{k=0}^{m-1} p^k s^{k+1} + qG_{T^{(m)}}(s) \sum_{k=1}^{m-1} (k+1)(ps)^k$$

$$= mp^m s^{m-1} + qsG'_{T^{(m)}}(s)\frac{1 - (ps)^m}{1 - ps} + (1 - p)G_{T^{(m)}}(s) \sum_{k=1}^{m-1} (k+1)(ps)^k.$$

Using the relations

$$mp^m + (1 - p) \sum_{k=1}^{m-1} (k+1)p^k = \frac{1 - p^m}{1 - p}, \qquad 0 \leqslant p < 1,$$

and $G_{T^{(m)}}(1) = \mathbb{P}\big(T^{(m)} < \infty\big) = 1$ from (3.10), we have

$$G'_{T^{(m)}}(1) = mp^m + (1-p)\sum_{k=0}^{m-1} p^k(k+1) + qG'_{T^{(m)}}(1)\sum_{k=0}^{m-1} p^k$$

$$= \frac{1-p^m}{1-p} + qG'_{T^{(m)}}(1)\sum_{k=0}^{m-1} p^k$$

$$= \frac{1-p^m}{1-p} + qG'_{T^{(m)}}(1)\frac{1-p^m}{1-p}$$

$$= \frac{1-p^m}{1-p} + G'_{T^{(m)}}(1)(1-p^m),$$

which yields (3.11) when $p \in [0,1)$. In case $p = 1$ and $q = 0$, we find $G'_{T^{(m)}}(1) = m$. □

For example, for an unbiased coin with $p = 1/2$ the mean time until the first winning streak of length $m \geqslant 1$ is

$$\mathbb{E}\big[T^{(m)}\big] = \sum_{k=1}^{m} \frac{1}{(1/2)^k} = \sum_{k=1}^{m} 2^k = 2\frac{1-2^m}{1-2} = 2(2^m - 1).$$

3.3 Synchronizing Automata

An *automaton* is given by a function

$$f : \{a, b\} \times \{0, 1, \ldots, n\} \longrightarrow \{0, 1, \ldots, n\}$$

and reads words of the form $a_1 a_2 \cdots a_m \in \mathcal{L}$ by producing a sequence $y_1, y_2, \ldots, y_m$ of integers starting from an initial y_0, *via* the following recursion:

$$y_1 := f(a_1, y_0), \quad y_2 := f(a_2, y_1), \quad y_3 := f(a_3, y_2), \ldots, \quad y_m := f(a_m, y_{m-1}).$$

Definition 3.6 A word $a_1 a_2 \cdots a_m \in \mathcal{L}$, $m \geqslant 1$, is said to *synchronize* the automaton f *to state* ⓝ if we have $y_m = $ ⓝ, where ⓝ is regarded as a *sink state*, also called an *accepting state*.

Example

Let $n = 5$, and consider the automaton given by the function

f	0	1	2	3	4	5
a	1	2	3	3	1	5
b	0	0	0	4	5	5

The automaton can be represented by the following graph.

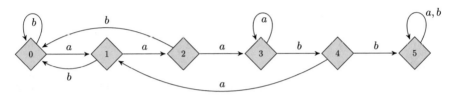

Definition 3.7 One says that the automaton f *recognizes* the language $\mathcal{L}$ if every word $a_1 a_2 \cdots a_m$ in $\mathcal{L}$, $m \geqslant 1$, synchronizes the automaton f *to state* $\textcircled{n}$, i.e. satisfies $y_m = \textcircled{n}$, starting from *any* initial state $\textcircled{y_0}$.

We note that the shortest word of the form "$a^l b^m$" which is synchronized to state $\textcircled{5}$ by the above automaton starting from *any* state is "$a^3 b^2$", with $l = 3$ and $m = 2$.

According to Definition 3.1, the set of words, or language, recognized by this automaton can be denoted by $\Sigma^* a^3 b^2 \Sigma^*$. An example of a five-letter word that *does not* synchronize the automaton when started from state $\textcircled{4}$ is by "$aabbb$".

Markovian Text Generator

In what follows, we "feed" the automaton with the *i.i.d.* sequence $(X_k)_{k \geqslant 1}$ of $\{a, b\}$-valued samples generated as in (3.2), i.e. such that

$$\mathbb{P}(X_k = a) = p \in (0, 1) \quad \text{and} \quad \mathbb{P}(X_k = b) = q = 1 - p, \quad k \geqslant 1.$$

This results into a random process $(Y_k)_{k \in \mathbb{N}}$ started at Y_0, with

$$Y_1 = f(X_1, Y_0), \ Y_2 = f(X_2, Y_1), \dots, Y_k = f(X_k, Y_{k-1}), \dots$$

is a Markov chain on the state space $\{0, 1, 2, 3, 4, 5\}$. Indeed, given Y_k, the distribution of $Y_{k+1} := f(X_{k+1}, Y_k)$ is independent of $Y_0, \dots, Y_{k-1}$. The graph of the chain $(Y_k)_{k \in \mathbb{N}}$ can be described as follows.

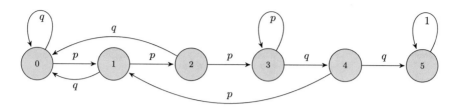

The chain $(Y_k)_{k\in\mathbb{N}}$ is reducible, its communicating classes are $\{0, 1, 2, 3, 4\}$ and $\{5\}$, and its transition matrix is given by

$$
[\,P_{i,j}\,]_{0\leqslant i,j\leqslant 5} =
\begin{bmatrix}
q & p & 0 & 0 & 0 & 0 \\
q & 0 & p & 0 & 0 & 0 \\
q & 0 & 0 & p & 0 & 0 \\
0 & 0 & 0 & p & q & 0 \\
0 & p & 0 & 0 & 0 & q \\
0 & 0 & 0 & 0 & 0 & 1
\end{bmatrix}.
\tag{3.12}
$$

3.4   Synchronization Times

Mean Synchronization Times

We compute the average time it takes until the automaton f of Sect. 3.3 becomes synchronized by the random words generated from $(X_k)_{k\geqslant 1}$, i.e. the mean time until the word "a^3b^2" is generated after starting from the initial state $Y_0 = \textcircled{0}$. Denoting by $h_5(k)$ the average time it takes to reach state $\textcircled{5}$ starting from state $k = 0, 1, 2, 3, 4, 5$, we first check that $h_5(4) = ph_5(0)$.

By first step analysis, we find the equations

$$
\begin{cases}
h_5(0) = 1 + qh_5(0) + ph_5(1) \\
h_5(1) = 1 + qh_5(0) + ph_5(2) \\
h_5(2) = 1 + qh_5(0) + ph_5(3) \\
h_5(3) = 1 + ph_5(3) + qh_5(4) \\
h_5(4) = 1 + ph_5(1) + qh_5(5) \\
h_5(5) = 0,
\end{cases}
$$

i.e.

$$
\begin{cases}
ph_5(0) = 1 + ph_5(1) \\
h_5(1) = 1 + qh_5(0) + ph_5(2) \\
h_5(2) = 1 + qh_5(0) + ph_5(3) \\
qh_5(3) = 1 + qh_5(4) = 1 + qph_5(0) \\
h_5(4) = 1 + ph_5(1) = ph_5(0) \\
h_5(5) = 0,
\end{cases}
$$

i.e.

$$
\begin{cases}
h_5(0) = \dfrac{1}{p} + \dfrac{1}{p^2} + \dfrac{1}{p^3} + h_5(3) \\
h_5(1) - \dfrac{1}{p^2} + \dfrac{1}{p^3} + h_5(3) \\
h_5(2) = \dfrac{1}{p^3} + h_5(3) \\
h_5(3) = \dfrac{1}{q} + ph_5(0) \\
h_5(4) = ph_5(0) \\
h_5(5) = 0,
\end{cases}
$$

i.e.

$$
\begin{cases}
h_5(0) = \dfrac{q(p^2 + p + 1) + p^3}{p^3 q^2} = \dfrac{1}{p^3 q^2} \\
h_5(1) = \dfrac{1}{p^3 q^2} - \dfrac{1}{p} \\
h_5(2) = \dfrac{1}{p^3 q^2} - \dfrac{1}{p} - \dfrac{1}{p^2} \\
h_5(3) = \dfrac{1}{q} + \dfrac{1}{p^2 q^2} \\
h_5(4) = \dfrac{1}{p^2 q^2} \\
h_5(5) = 0.
\end{cases}
$$

Synchronization Probabilities

For another example, let $n = 4$ and consider the automaton f defined by

f	0	1	2	3	4
a	0	2	2	1	4
b	0	0	3	4	4

This automaton has two sink states ⓪ and ④, and its graph is given as follows:

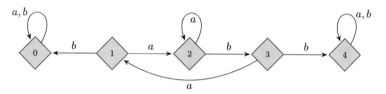

We note that the unique shortest word that synchronizes this automaton to state ⓪ after starting from all states 1, 2, 3 is "*abab*". Similarly, the unique shortest word that synchronizes to state ④ starting from all states 1, 2, 3 is "*aabb*".

The random process $(Y_k)_{k\in\mathbb{N}}$ started at Y_0, with

$$Y_1 = f(X_1, Y_0), \ Y_2 = f(X_2, Y_1), \ldots, Y_k = f(X_k, Y_{k-1}), \ldots$$

is a Markov chain with transition matrix

$$\left[P_{i,j} \right]_{0\leqslant i,j\leqslant 4} = \begin{bmatrix} 1 & 0 & 0 & 0 & 0 \\ q & 0 & p & 0 & 0 \\ 0 & 0 & p & q & 0 \\ 0 & p & 0 & 0 & q \\ 0 & 0 & 0 & 0 & 1 \end{bmatrix}$$

on the state space $\{0, 1, 2, 3, 4\}$, with the following graph.

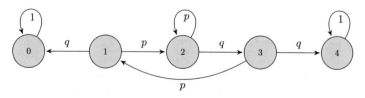

The following result is an application of Proposition 1.2 with the boundary set $\mathcal{A} = \{0, 4\}$.

Proposition 3.8 *The probability that the first synchronized word is "abab" when the automaton is started from state* ① *is* $p^2/(1+p)$.

Proof We note that synchronization may here occur through state ⓪ or through state ④. Denoting by $g_0(k)$ the probability that state ⓪ is reached first starting from state $k = 0, 1, 2, 3, 4$, we have the equations

$$\begin{cases} g_0(0) = 1 \\ g_0(1) = qg_0(0) + pg_0(2) = q + pg_0(2) \\ g_0(2) = pg_0(2) + qg_0(3) \\ g_0(3) = pg_0(1) + qg_0(4) = pg_0(1) \\ g_0(4) = 0, \end{cases}$$

i.e.

$$\begin{cases} g_0(0) = 1 \\ g_0(1) = q + pg_0(2) \\ g_0(2) = pg_0(2) + qg_0(3) = pg_0(2) + qpg_0(1) \\ g_0(3) = pg_0(1) \\ g_0(4) = 0, \end{cases}$$

i.e.

$$\begin{cases} g_0(0) = 1 \\ g_0(1) = q + p^2g_0(1) \\ g_0(2) = pg_0(1) \\ g_0(3) = pg_0(1) \\ g_0(4) = 0, \end{cases}$$

or

$$\begin{cases} g_0(0) = 1 \\ g_0(1) = \dfrac{q}{1-p^2} = \dfrac{1}{1+p} \\ g_0(2) = \dfrac{pq}{1-p^2} = \dfrac{p}{1+p} \\ g_0(3) = \dfrac{pq}{1-p^2} = \dfrac{p}{1+p} \\ g_0(4) = 0. \end{cases}$$

Now, starting from state ① one may move directly to state ⓪ with probability q, in which case the first synchronized word is "b", not "abab". For this reason we need to subtract q from $g_0(1)$, and the probability that the first synchronized word is "abab" starting from state $\textcircled{1}$ is

$$\frac{1}{1+p} - (1-p) = \frac{p^2}{1+p}.$$

$\square$

Note that the above computations apply only when $p \in [0, 1)$. In case $p = 1$ the problem admits a trivial solution since the word "abab" will never occur.

The probability $g_0(1)$ can also be computed by pathwise analysis and a geometric series, as $g_0(1) = 1 - g_4(1)$, with $g_4(1) = g_3(2)g_4(3)$, where

$$g_3(2) = q \sum_{k \geqslant 0} p^k = \frac{q}{1-p} = 1$$

and

$$g_4(3) = q \sum_{k \geqslant 0} p^{2k} = \frac{q}{1-p^2},$$

hence

$$g_0(1) = 1 - pq \sum_{k \geqslant 0} p^{2k} = 1 - \frac{pq}{1-p^2} = 1 - \frac{p}{1+p} = \frac{1}{1+p}.$$

The averages times until the automaton is synchronized by the word "*abab*" or by the word "*aabb*" can be similarly computed by first step analysis.

Notes

See e.g. Volkov (2008) and Gusev (2014) for further reading.

Exercises

Exercise 3.1 Consider a sequence $(X_n)_{n \geqslant 1}$ of independent Bernoulli random variables with the distribution

$$\mathbb{P}(X_n = a) = p, \qquad \mathbb{P}(X_n = b) = q, \qquad n \geqslant 1,$$

where $p \in (0, 1]$ and $q := 1 - p$. Let $T^{(m)}$ denote the time of the first appearance of m consecutive a's in $(X_n)_{n \geq 1}$, with e.g. $T^{(4)} = 8$ in the following sequence:

$$(b, a, a, b, \overbrace{a, a, a, a}^{4 \text{ times}}, b, a, a, b, \ldots).$$
$$\begin{array}{ccccccccc} & \uparrow & \uparrow & \uparrow & \uparrow & \uparrow & \uparrow & \uparrow & \uparrow \\ & 1 & 2 & 3 & 4 & 5 & 6 & 7 & 8 \end{array}$$

(a) By first step analysis, find an equation satisfied by $\mathbb{E}[T^{(m)}]$.
(b) Compute the mean time $\mathbb{E}[T^{(m)}]$ until we encounter an m-winning streak, for all $m \geq 1$.

Hint. We have

$$\sum_{k=1}^{m} k p^{k-1} = \frac{\partial}{\partial p} \sum_{k=0}^{m} p^k = \frac{\partial}{\partial p} \left(\frac{1 - p^{m+1}}{1 - p} \right) = \frac{1 - (m+1)p^m + mp^{m+1}}{(1 - p)^2}.$$

Exercise 3.2 Consider a sequence $(X_n)_{n \in \mathbb{Z}}$ of independent Bernoulli random variables with the distribution

$$\mathbb{P}(X_n = a) = p, \qquad \mathbb{P}(X_n = b) = q, \qquad n \in \mathbb{Z},$$

where $p \in (0, 1]$ and $q := 1 - p$, and let $m \geq 1$ be a fixed integer.

For $n \geq 0$, we let Z_n denote the smallest of m and the number of "a" having appeared up to time n since the last occurrence of "b" in the sequence $(X_k)_{k \leq n}$. For example, in the sequence

$$(a, b, a, a, a, a, a, a, b, a, a, a, a, b, a, a, b, \ldots),$$
$$\begin{array}{ccccccccccccccccc} \uparrow & \uparrow & \uparrow & \uparrow & \uparrow & \uparrow & \uparrow & \uparrow & \uparrow & \uparrow & \uparrow & \uparrow & \uparrow & \uparrow & \uparrow & \uparrow & \uparrow \\ -5 & -4 & -3 & -2 & -1 & 0 & 1 & 2 & 3 & 4 & 5 & 6 & 7 & 8 & 9 & 10 & 11 \end{array}$$

we have

$$Z_0 = 4, \ Z_1 = 5, \ Z_2 = 6, \ Z_3 = 0, \ Z_4 = 1, \ Z_5 = 2, \ Z_6 = 3,$$
$$Z_7 = 4, \ Z_8 = 0, \ Z_9 = 1.$$

(a) Show that $(Z_n)_{n \geq 0}$ is a Markov chain, give its state space and transition matrix P.
(b) Compute the mean hitting time $\mathbb{E}[T_m \mid Z_0 = l]$ of state m by the chain $(Z_n)_{n \geq 0}$ after starting from $Z_0 = l$, for $l \in \{0, 1, \ldots, m\}$.
(c) Give the expected value of the time $T^{(m)}$ of the first appearance of m consecutive "a" in the sequence $(X_n)_{n \geq 1}$, and recover the expected value of $T^{(m)}$ obtained in Question (b) of Exercise 3.1.

For example, taking $m := 4$ we have $T^{(4)} = 8$ in the following sequence:

$$
(\underbrace{b}_{\substack{\uparrow \\ -3}}, \underbrace{a}_{\substack{\uparrow \\ -2}}, a, b, \overbrace{a, a, a, a}^{4 \text{ times}}, b, a, a, b, \ldots).
$$

Problem 3.3 Pattern recognition. Consider a sequence $(X_n)_{n \geqslant 0}$ of *i.i.d.* Bernoulli random variables taking values in a two-letter alphabet $\{a, b\}$, with

$$
\mathbb{P}(X_n = a) = p \quad \text{and} \quad \mathbb{P}(X_n = b) = q = 1 - p, \quad n \geqslant 0,
$$

with $0 < p < 1$, and the discrete-time process $(Z_n)_{n \geqslant 1}$ defined by

$$
Z_n := (X_{n-1}, X_n), \qquad n \geqslant 1.
$$

(a) Argue that $(Z_n)_{n \geqslant 1}$ is a Markov chain with four possible states (or words) $\{aa, ab, ba, bb\}$, and write down its 4×4 transition matrix.

(b) Let

$$
\tau_{ab} = \inf\{n \geqslant 1 \ : \ Z_n = (a, b)\}
$$

denote the first time of appearance of the pattern "*ab*" in the sequence $(X_0, X_1, X_2, \ldots)$. Give the value of

$$
G_{ab}(s) := \mathbb{E}\big[s^{\tau_{ab}} \, \big| \, Z_1 = (a, b)\big], \qquad -1 < s < 1.
$$

(c) Consider the probability generating functions

$$
G_{aa}(s) := \mathbb{E}\big[s^{\tau_{ab}} \, \big| \, Z_1 = (a, a)\big], \quad \text{and} \quad G_{ba}(s) := \mathbb{E}\big[s^{\tau_{ab}} \, \big| \, Z_1 = (b, a)\big],
$$

$-1 < s < 1$. Using first step analysis, complete the system of equations

$$
\begin{cases} G_{aa}(s) = ps\,G_{aa}(s) + qs\,G_{ab}(s), \\[2mm] G_{ba}(s) = \quad ? \quad + \quad ? \end{cases} \tag{3.13}
$$

(d) Compute $G_{aa}(s)$ and $G_{ba}(s)$ by solving the system (3.13).

(e) Using probability generating functions, compute the averages

$$
\mathbb{E}[\tau_{ab} \mid Z_1 = (a, a)] \quad \text{and} \quad \mathbb{E}[\tau_{ab} \mid Z_1 = (b, a)].
$$

(f) Find the average time it takes until we encounter the pattern "*ab*" in the sequence $(X_0, X_1, X_2, \ldots)$ started with $X_0 = a$.

Exercise 3.4 Consider the probabilistic automaton g defined by

g	0	1	2	3	4
a	0	2	2	1	4
b	0	0	3	4	4

This automaton has two sink states ⓪ and ④, and its graph is given as follows:

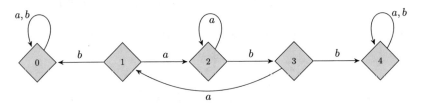

(a) Find the shortest word that synchronizes this automaton to state ④ after starting from any of the states 1, 2, 3.
(b) Consider the $\{a, b\}$-valued two-state Markov chain $(X_n)_{n\geqslant 0}$ with *transition probability matrix*

$$P = \begin{bmatrix} 1/2 & 1/2 \\ 1/2 & 1/2 \end{bmatrix}$$

and the Markov chain on $(Z_k)_{k\in\mathbb{N}}$ the state space $\{0, 1, 2, 3, 4\}$ started at Z_0, with

$$Z_1 = g(X_1, Z_0), \ Z_2 = g(X_2, Z_1), \ldots, \ Z_k = g(X_k, Z_{k-1}), \ldots$$

Draw the graph of the chain $(Z_k)_{k\in\mathbb{N}}$ and write down its transition probability matrix.
(c) Find the probability that the first synchronized word is "aabb" when the automaton is started from state ①.

Exercise 3.5 Consider the probabilistic automaton g defined by

g	0	1	2	3	4
a	0	2	2	4	4
b	0	0	3	1	4

This automaton has two sink states 0 and 4, and its graph is given as follows:

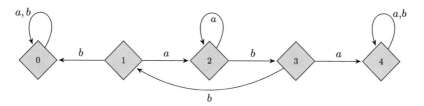

(a) Find the unique shortest word that synchronizes this automaton to state 4 after starting from any of the states $1, 2, 3$.

(b) We assume that letters are generated from an $\{a, b\}$-valued two-state Markov chain $(X_n)_{n \geq 0}$ with the *transition probability matrix*

$$P = \begin{bmatrix} 1/2 & 1/2 \\ 1/2 & 1/2 \end{bmatrix}.$$

Find the probability that the first synchronized word is "aba" when the automaton is started from state 1.

Exercise 3.6 Using Proposition 3.4 and Relation (A.7), compute the probability distribution of $T^{(m)}$ on $\{m, m + 1, \ldots\}$.

Chapter 4
Random Walks and Recurrence

This chapter reviews the recurrence and transience properties of multidimensional random walks, and considers the calculation of hitting probabilities and mean hitting times in more sophisticated examples such as reflected and conditioned random walks. Those results will be applied to the study of random walks in cookie environment, or excited random walks, in Chap. 5.

4.1 Distribution and Hitting Times

Let $\{e_1, e_2, \ldots, e_d\}$ denote the canonical basis of $\mathbb{R}^d$, i.e.

$$e_k = (0, \ldots, 0, \underset{\underset{\mathbf{k}}{\uparrow}}{1}, 0, \ldots, 0), \qquad k = 1, 2, \ldots, d.$$

The unrestricted $\mathbb{Z}^d$-valued random walk $(S_n)_{n \geqslant 0}$, also called the *Bernoulli* random walk, is defined by $S_0 = 0$ and

$$S_n = \sum_{k=1}^{n} X_k = X_1 + \cdots + X_n, \qquad n \geqslant 0,$$

started at

$$S_0 = \vec{0} = \underbrace{(0, \ldots, 0)}_{d \text{ times}},$$

where the random walk *increments*

$$X_n \in \{e_1, e_2, \ldots, e_d, -e_1, -e_2, \ldots, -e_d\}, \qquad n \geqslant 1,$$

form an *independent and identically distributed (i.i.d.)* family $(X_n)_{n \geqslant 1}$ of random variables with distribution

$$\mathbb{P}(X_n = e_k) = p_k, \qquad \mathbb{P}(X_n = -e_k) = q_k, \qquad k = 1, 2, \ldots, d,$$

such that

$$\sum_{k=1}^{d} p_k + \sum_{k=1}^{d} q_k = 1.$$

One-Dimensional Random Walk

When $d = 1$, the distribution of S_{2n} is given by

$$\mathbb{P}(S_{2n} = 2k \mid S_0 = 0) = \binom{2n}{n+k} p^{n+k} q^{n-k}, \qquad -n \leqslant k \leqslant n, \tag{4.1}$$

and we note that in an even number of time steps, $(S_n)_{n \geqslant 0}$ can only reach an even state in $\mathbb{Z}$ starting from $\textcircled{0}$. Similarly, in an odd number of time steps, $(S_n)_{n \geqslant 0}$ can only reach an odd state in $\mathbb{Z}$ starting from $\textcircled{0}$. In Fig. 4.1 we enumerate the $120 = \binom{10}{7} = \binom{10}{3}$ possible paths corresponding to $n = 5$ and $k = 2$.

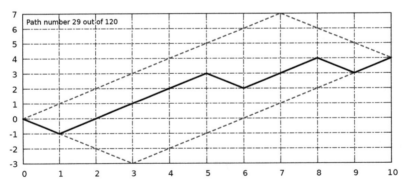

Fig. 4.1 Graph of $120 = \binom{10}{7} = \binom{10}{3}$ paths linking $(0, 0)$ to $(10, 4)$

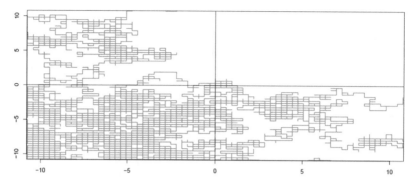

Fig. 4.2 Two-dimensional random walk

Two-Dimensional Random Walk

When $d = 2$ the random walk can return to state $\vec{0}$ in $2n$ time steps by

- k forward steps in the direction e_1,
- k backward steps in the direction $-e_1$,
- $n - k$ forward steps in the direction e_2,
- $n - k$ backward steps in the direction $-e_2$,

where k ranges from 0 to $2n$. Figure 4.2 can be plotted using the ℝ code_05_page_87.R.

For each $k = 0, 1, \ldots, n$ the number of ways to arrange those four types of moves among $2n$ time steps is the multinomial coefficient

$$\binom{2n}{k, k, n - k, n - k} = \frac{(2n!)}{k!k!(n - k)!(n - k)!},$$

hence, since every sequence of $2n$ moves occur with the same probability $(1/4)^{2n}$, by summation over $k = 0, 1, \ldots, n$ we find

$$\mathbb{P}\big(S_{2n} = \vec{0} \mid S_0 = \vec{0}\big) = \sum_{k=0}^{n} \frac{(2n!)}{(k!)^2((n - k)!)^2}(p_1 q_1)^k (p_2 q_2)^{n-k}$$

$$= \frac{(2n)!}{(n!)^2} \sum_{k=0}^{n} \binom{n}{k}^2 (p_1 q_1)^k (p_2 q_2)^{n-k}. \qquad (4.2)$$

Multidimensional Random Walk

Given $i_1, i_2, \ldots, i_d \in \mathbb{N}$, we count all paths starting from $\vec{0}$ and returning to $\vec{0}$ via i_k "forward" steps in the direction e_k and i_k "backward" steps in the direction $-e_k$, $k = 1, 2, \ldots, d$.

In order to come back to $\vec{0}$ we need to take i_1 forward steps in the direction e_1 and i_1 backward steps in the direction e_1, and similarly for $i_2, \ldots, i_d$. The number of ways to arrange such paths is given by the multinomial coefficients

$$\binom{2n}{i_1, i_1, i_2, i_2, \ldots, i_d, i_d} = \frac{(2n)!}{(i_1!)^2 \cdots (i_d!)^2},$$

and by summation over all possible indices $i_1, i_2, \ldots, i_d \geqslant 0$ satisfying $i_1 + \cdots + i_d = n$ and multiplying by the probability $(1/(2d))^{2n}$ of each path, we find

$$\mathbb{P}\big(S_{2n} = \vec{0}\big) = \sum_{\substack{i_1 + \cdots + i_d = n \\ i_1, i_2, \ldots, i_d \geqslant 0}} \binom{2n}{i_1, i_1, i_2, i_2, \ldots, i_d, i_d} \prod_{k=1}^{d} (p_k q_k)^{i_k}$$

$$= \sum_{\substack{i_1 + \cdots + i_d = n \\ i_1, i_2, \ldots, i_d \geqslant 0}} \frac{(2n)!}{(i_1!)^2 \cdots (i_d!)^2} \prod_{k=1}^{d} (p_k q_k)^{i_k}. \qquad (4.3)$$

Figure 4.3 can be plotted using the ®R code_06_page_88.R.

Fig. 4.3 Three-dimensional
random walk

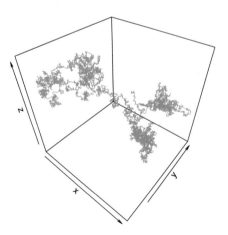

Return Times

We let

$$T_0^r := \inf\{n \geqslant 1 \ : \ S_n = 0\}$$

denote the first return time to $\textcircled{0}$ of the one-dimensional random walk $(S_n)_{n \geqslant 0}$, as illustrated in the $\textcircled{R}$ code_07_page_89.R. The proof of the following proposition relies on the *reflection principle*.

Proposition 4.1 *The probability distribution* $\mathbb{P}(T_0^r = n \mid S_0 = 0)$ *of the first return time* T_0^r *to* $\textcircled{0}$ *is given by*

$$\mathbb{P}\big(T_0^r = 2n \mid S_0 = 0\big) = \frac{1}{2n-1}\binom{2n}{n}(pq)^n, \qquad n \geqslant 1,$$

with $\mathbb{P}(T_0^r = 2n + 1 \mid S_0 = 0) = 0, n \geqslant 0.$

Proof

(a) We first note that the number of paths joining $S_0 = 0$ to $S_{2n} = 0$ without hitting $\textcircled{0}$ can be split into the sets of paths joining $S_1 = 1$ to $S_{2n-1} = 1$ without hitting $\textcircled{0}$ on the one hand, and the sets of paths joining $S_1 = -1$ to $S_{2n-1} = -1$ without hitting $\textcircled{0}$ on the other hand. According to the graph of Fig. 4.4, to each blue path joining $S_1 = 1$ to $S_{2n-2} = 1$ *without* hitting $\textcircled{0}$ between time 1 and time $2n - 1$, we can associate a unique red path joining $S_1 = -1$ to $S_{2n-2} = -1$ *without* hitting $\textcircled{0}$.

(b) On the graph of Fig. 4.5, every blue path joining $S_1 = 1$ to $S_{2n-1} = 1$ by hitting $\textcircled{0}$ is associated to a unique red path joining $S_1 = 1$ to $S_{2n-1} = -1$, which is the reflection of the blue path starting at the first time τ it hits $\textcircled{0}$. As in (4.1),

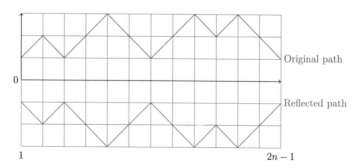

Fig. 4.4 Random walk and reflected path

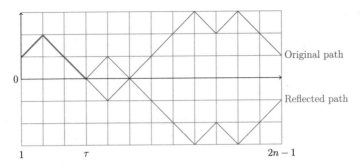

Fig. 4.5 Random walk and reflected path

the count of such paths is

$$\binom{2n-2}{n-2} = \binom{2n-2}{n}.$$

Knowing that, by (4.1), the total count of paths joining $S_0 = 1$ to $S_{2n} = 1$ is $\binom{2n-2}{n-1}$, we find that the number of paths joining $S_1 = 1$ to $S_{2n-2} = 1$ *without* crossing $\textcircled{0}$ between time 1 and time $2n - 1$ is

$$\binom{2n-2}{n-1} - \binom{2n-2}{n-2} = \frac{(2n-2)!}{(n-1)!(n-1)!} - \frac{(2n-2)!}{(n-2)!n!}$$

$$= \frac{(n^2 - n(n-1))(2n-2)!}{n!n!}$$

$$= \frac{(2n-2)!}{(n-1)!n!}.$$

Adding the number of paths joining $S_1 = 1$ to $S_{2n-2} = 1$ *without* crossing $\textcircled{0}$ between time 1 and time $2n - 1$ to the number of paths joining $S_1 = -1$ to $S_{2n-2} = -1$ *without* crossing $\textcircled{0}$ between time 1 and time $2n - 1$, we get the total to the number of paths joining $S_0 = 0$ to $S_{2n} = 0$ *without* crossing $\textcircled{0}$, between time 0 and time $2n$, as follows:

$$2 \times \frac{(2n-2)!}{(n-1)!n!} = \frac{2n(2n-2)!}{n!n!} = \frac{1}{2n-1}\binom{2n}{n}.$$

$\square$

Let

$$G_{T_0^r} : [-1, 1] \longrightarrow \mathbb{R}$$

$$s \longmapsto G_{T_0^r}(s)$$

denote the Probability Generating Function (PGF) of the random variable T_0^r, defined by

$$G_{T_0^r}(s) := \mathbb{E}\big[s^{T_0^r} \mathbb{1}_{\{T_0^r < \infty\}} \,\big|\, S_0 = 0\big] = \sum_{n \geqslant 0} s^n \mathbb{P}\big(T_0^r = n \,\big|\, S_0 = 0\big),$$

$-1 \leqslant s \leqslant 1$, cf. (A.1). Recall that the knowledge of $G_{T_0^r}(s)$ provides allows us to recover the finite return time probability

$$\mathbb{P}\big(T_0^r < \infty \mid S_0 = 0\big) = \mathbb{E}\big[\mathbb{1}_{\{T_0^r < \infty\}} \,\big|\, S_0 = 0\big] = G_{T_0^r}(1),$$

and the return time expectation

$$\mathbb{E}\big[T_0^r \mathbb{1}_{\{T_0^r < \infty\}} \,\big|\, S_0 = 0\big] = \sum_{n \geqslant 1} n \mathbb{P}\big(T_0^r = n \,\big|\, S_0 = 0\big) = G'_{T_0^r}(1^-).$$

The following result is a consequence of Proposition 4.1, and can be obtained in Mathematica via the command Sum[Bin[2*n,n]*(p*q*s²)ⁿ/(2*n-1),{n,1,Infinity}].

Proposition 4.2 *The probability generating function $G_{T_0^r}$ of the first return time T_0^r to* ⓪ *is given by*

$$G_{T_0^r}(s) = 1 - \sqrt{1 - 4pqs^2}, \qquad 4pqs^2 < 1. \tag{4.4}$$

Proof By Proposition 4.1, the probability distribution $\mathbb{P}\big(T_0^r = n \mid S_0 = 0\big)$ of the first return time T_0^r to ⓪ is given by

$$\mathbb{P}(T_0^r = 2k \mid S_0 = 0) = \frac{1}{2k - 1}\binom{2k}{k}(pq)^k, \qquad k \geqslant 1,$$

with $\mathbb{P}\big(T_0^r = 2k + 1 \mid S_0 = 0\big) = 0$, $k \in \mathbb{N}$. By applying a Taylor expansion to $s \longmapsto 1 - (1 - 4pqs^2)^{1/2}$ in (4.4), we get

$$G_{T_0^r}(s) = \sum_{n \geqslant 0} s^n \mathbb{P}\big(T_0^r = n \mid S_0 = 0\big)$$

$$= \sum_{k \geqslant 1} s^{2k} \mathbb{P}\big(T_0^r = 2k \mid S_0 = 0\big)$$

$$= \sum_{k \geqslant 1} \frac{s^{2k}}{2k-1} \binom{2k}{k} (pq)^k$$

$$= \sum_{k \geqslant 1} \frac{s^{2k}}{k!} \frac{1}{2k-1} \frac{1 \times 2 \times \cdots \times (2k-1) \times (2k)}{1 \times 2 \times \cdots \times (k-1) \times k} (pq)^k$$

$$= \sum_{k \geqslant 1} \frac{s^{2k}}{k!} \frac{1}{2k-1} 1 \times 3 \times 5 \times \cdots \times (2k-3) \times (2k-1)(2pq)^k$$

$$= \frac{1}{2} \sum_{k \geqslant 1} s^{2k} \frac{(4pq)^k}{k!} \left(1 - \frac{1}{2}\right) \times \cdots \times \left(k - 1 - \frac{1}{2}\right)$$

$$= 1 - \sum_{k \geqslant 0} \frac{1}{k!} (-4pqs^2)^k \left(\frac{1}{2} - 0\right) \left(\frac{1}{2} - 1\right) \times \cdots \times \left(\frac{1}{2} - (k-1)\right)$$

$$= 1 - (1 - 4pqs^2)^{1/2},$$

where we used the Taylor expansion

$$(1 + x)^\alpha = \sum_{k \geqslant 0} \frac{x^k}{k!} \alpha(\alpha - 1) \times \cdots \times (\alpha - (k-1))$$

for $\alpha = 1/2$. $\square$

The distribution

$$\mathbb{P}(T_0^r = 2k \mid S_0 = 0) = \frac{(4pq)^k}{k!} \frac{1}{2} \left(1 - \frac{1}{2}\right) \times \cdots \times \left(k - 1 - \frac{1}{2}\right)$$

$$= \frac{(4pq)^k}{2k!} \prod_{m=1}^{k-1} \left(m - \frac{1}{2}\right)$$

$$= \frac{1}{2k-1} \binom{2k}{k} (pq)^k, \qquad k \geqslant 1,$$

can be recovered from the relation

$$\mathbb{P}(T_0^r = n \mid S_0 = 0) = \frac{1}{n!} \frac{\partial^n}{\partial s^n} G_{T_0^r}(s)_{|s=0}, \qquad n \geqslant 0.$$

Proposition 4.3 *The probability that the first return to* $\textcircled{0}$ *occurs within a finite time is*

$$\mathbb{P}(T_0^r < \infty \mid S_0 = 0) = 2 \min(p, q), \qquad (4.5)$$

and we have

$$\mathbb{P}(T_0^r = \infty \mid S_0 = 0) = |2p - 1| = |p - q|. \tag{4.6}$$

Proof We have

$$\mathbb{P}(T_0^r < \infty \mid S_0 = 0) = \mathbb{E}\left[\mathbb{1}_{\{T_0^r < \infty\}} \mid S_0 = 0\right] = \mathbb{E}\left[1^{T_0^r} \mathbb{1}_{\{T_0^r < \infty\}} \mid S_0 = 0\right]$$

$$= G_{T_0^r}(1) = 1 - \sqrt{1 - 4pq}$$

$$= 1 - |2p - 1| = 1 - |p - q| = \begin{cases} 2q, & p \geqslant 1/2, \\ \\ 2p, & p \leqslant 1/2, \end{cases}$$

$$= 2\min(p, q),$$

hence

$$\mathbb{P}(T_0^r = \infty \mid S_0 = 0) = 1 - \mathbb{P}(T_0^r < \infty \mid S_0 = 0) = |2p - 1| = |p - q|.$$

which can be obtained in Mathematica via the command

$$\text{Sum[Bin[2*n,n]*(p*q)}^n\text{/(2*n-1),\{n,1,Infinity\}]}.$$

or using the following Python code:

```
from sympy import *
import sympy as sp
k = sp.Symbol("k");p = sp.Symbol("p"); q = sp.Symbol("q")
prob=summation(p**k*q**k*factorial(2*k)/factorial(k)**2/(2*k-1), (k, 1, oo))
simplify(prob.args[0][0])
```

□

We make the following comments.

(i) In the non-symmetric case $p \neq q$, Relation (4.5) shows that

$$\mathbb{P}(T_0^r < \infty \mid S_0 = 0) < 1 \qquad \text{and} \qquad \mathbb{P}(T_0^r = \infty \mid S_0 = 0) > 0.$$

In addition, by (4.6), the time T_0^r needed to return to state ⓪ is infinite with probability

$$\mathbb{P}(T_0^r = \infty \mid S_0 = 0) = |p - q| > 0,$$

hence

$$\mathbb{E}[T_0^r \mid S_0 = 0] = \infty. \tag{4.7}$$

i.e. the symmetric random walk is *null recurrent* according to Definition 1.19.

Starting from $S_0 = k \geqslant 1$, the mean hitting time of state $\textcircled{0}$ equals

$$\mathbb{E}[T_0^r \mid S_0 = k] = \begin{cases} \infty & \text{if } q \leqslant p, \\[2mm] \dfrac{k}{q-p} & \text{if } q > p, \end{cases} \tag{4.8}$$

see Exercise 3.2 in Privault (2018).

(ii) In the symmetric case $p = q = 1/2$ (or fair game) $p = q = 1/2$ we find that

$$\mathbb{P}(T_0^r < \infty \mid S_0 = 0) = 1 \qquad \text{and} \qquad \mathbb{P}(T_0^r = \infty \mid S_0 = 0) = 0,$$

i.e. the symmetric random walk is *recurrent*, as it returns to $\textcircled{0}$ with probability one and has a single communicating class, see Corollary 1.13. In addition, we have $\mathbb{P}(T_0^r < \infty \mid S_0 = 0) = 1$ and

$$\mathbb{E}[T_0^r \mid S_0 = 0] = \mathbb{E}[T_0^r \mathbb{1}_{\{T_0^r < \infty\}} \mid S_0 = 0] = G'_{T_0^r}(1^-) = \infty, \tag{4.9}$$

i.e. the symmetric random walk is *null recurrent* according to Definition 1.19.

This yields an example of a random variable T_0^r which is almost surely finite, while its expectation is infinite as in the St. Petersburg paradox which is illustrated in the ®️ code_09_page_94.R included in the supplementary material.

This shows how even a fair game can be risky when the player's initial wealth is negative, as it will take on average an infinite time to recover the losses.

From Proposition 4.1, we can also compute a conditional mean return time to $\textcircled{0}$ as

$$\mathbb{E}[T_0^r \mathbb{1}_{\{T_0^r < \infty\}} \mid S_0 = 0] = \sum_{n \geqslant 1} n \mathbb{P}(T_0^r = n \mid S_0 = 0)$$

$$= 2 \sum_{k \geqslant 1} k \mathbb{P}(T_0^r = 2k \mid S_0 = 0)$$

$$= 2 \sum_{k \geqslant 1} \frac{k}{2k-1} \binom{2k}{k} (pq)^k$$

$$= \frac{4pq}{|p-q|},$$

which can be computed by the following Python code:

```
1  from sympy import *
   import sympy as sp
3  n = sp.Symbol("n");p = sp.Symbol("p"); q = sp.Symbol("q")
   expectation=summation(2*n*p**n*q**n*factorial(2*n)/factorial(n)**2/(2*n-1), (n, 1,
       oo))
5  expectation.args[1].args[0][0]
```

When $p = q = 1/2$, we find

$$\mathbb{E}\big[T_0^r \mathbb{1}_{\{T_0^r < \infty\}} \,\big|\, S_0 = 0\big] = \sum_{k \geqslant 1} \frac{2k}{2k-1} \binom{2k}{k} \frac{1}{2^{2k}}. \tag{4.10}$$

Remark 4.4 By Stirling's approximation $k! \simeq (k/e)^k \sqrt{2\pi k}$ as k tends to infinity, we have

$$\frac{2k}{2k-1} \frac{1}{2^{2k}} \binom{2k}{k} = \frac{2k}{2k-1} \frac{(2k)!}{2^{2k}(k!)^2} \underset{k \to \infty}{\simeq} \frac{1}{\sqrt{\pi k}},$$

from which (4.10) recovers (4.9) by the limit comparison test.

The probability of hitting state $\textcircled{0}$ in finite time starting from any state $\textcircled{k}$ with $k \geqslant 1$ is given by

$$\mathbb{P}\big(T_0^r < \infty \mid S_0 = k\big) = \min\left(1, \left(\frac{q}{p}\right)^k\right), \qquad k \geqslant 1, \tag{4.11}$$

i.e.

$$\mathbb{P}\big(T_0^r = \infty \mid S_0 = k\big) = \mathrm{Max}\left(0, 1 - \left(\frac{q}{p}\right)^k\right), \qquad k \geqslant 1.$$

Using the independence of increments of the random walk $(S_n)_{n \geqslant 0}$, one can also show that the probability generating function of the first passage time

$$T_k = \inf\{n \geqslant 0 \,:\, S_n = k\}$$

to any level $k \geqslant 1$ is given by

$$G_{T_k}(s) = \left(\frac{1 - \sqrt{1 - 4pqs^2}}{2qs}\right)^k, \qquad 4pqs^2 < 1, \quad q \leqslant p, \tag{4.12}$$

from which the distribution of T_k can be computed given the series expansion of $G_{T_k}(s)$.

Fig. 4.6 Sample path of the
random walk $(S_n)_{n \geqslant 0}$

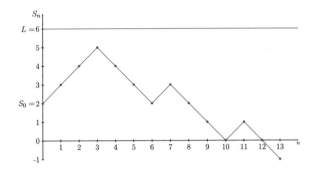

4.2 Recurrence of Symmetric Random Walks

The question of recurrence of the d-dimensional symmetric random walk has been
first solved in Pólya (1921). The treatment proposed in this section is based on
Champion et al. (2007). We consider the symmetric $\mathbb{Z}^d$-valued random walk

$$S_n = X_1 + \cdots + X_n, \qquad n \geqslant 0,$$

started at $S_0 = \vec{0} = (0, 0, \ldots, 0)$, where $(X_n)_{n \geqslant 1}$ is a sequence of independent
uniformly distributed random variables

$$X_n \in \{e_1, e_2, \ldots, e_d, -e_1, -e_2, \ldots, -e_d\}, \qquad n \geqslant 1,$$

with distribution

$$\mathbb{P}(X_n = e_k) = \mathbb{P}(X_n = -e_k) = \frac{1}{2d}, \qquad k = 1, 2, \ldots, d.$$

Let

$$T_{\vec{0}}^r := \inf\{n \geqslant 1 \ : \ S_n = \vec{0}\}$$

denote the time of first return[1] to $\vec{0} = (0, 0, \ldots, 0)$ of the random walk $(S_n)_{n \geqslant 0}$
started at $\vec{0}$, with the convention $\inf \emptyset = +\infty$, see Fig. 4.6.

Definition 4.5 The random walk $(S_n)_{n \geqslant 0}$ is said to be *recurrent* if $\mathbb{P}\big(T_{\vec{0}}^r < \infty\big) = 1$.

---

[1] Recall that the notation "inf" stands for "infimum", meaning here the smallest $n \geqslant 0$ such that
$S_n = 0$, with $T_0^r = +\infty$ if no such $n \geqslant 0$ exists.

Recurrence of the One-Dimensional Random Walk

When $d = 1$ we can now compute $\mathbb{P}(S_{2n} = 0)$, $n \geqslant 1$, and deduce that the one-dimensional random walk is *recurrent*, i.e. we have $\mathbb{P}(T_0^r < \infty) = 1$. For this, we will use Stirling's approximation $n! \simeq (n/e)^n \sqrt{2\pi n}$ as n tends to infinity. When $d = 1$, we have

$$[P^{2n}]_{0,0} = \mathbb{P}(S_{2n} = 0) = \frac{1}{2^{2n}} \binom{2n}{n} = \frac{(2n)!}{2^{2n}(n!)^2} \simeq_{n\to\infty} \frac{1}{\sqrt{\pi n}},$$

by Stirling's approximation, hence

$$\sum_{n\geqslant 0} [P^{2n}]_{0,0} = \infty.$$

and by Corollaries 1.12 or 4.12 below, we conclude that $\mathbb{P}(T_0^r < \infty) = 1$, i.e. we recover the fact that the one-dimensional symmetric random walk is *recurrent*.

Recurrence of the Two-Dimensional Random Walk

Proposition 4.6 *When $d = 2$ and $p_1 = q_1 = p_2 = q_2 = 1/4$, the two-dimensional symmetric random walk is* recurrent, *i.e. we have $\mathbb{P}(T_0^r < \infty) = 1$.*

Proof Recall that when $d = 2$, by (4.2) we have

$$[P^{2n}]_{\vec{0},\vec{0}} = \mathbb{P}(S_{2n} = \vec{0})$$

$$= \left(\frac{1}{4}\right)^{2n} \sum_{k=0}^{n} \frac{(2n)!}{(k!)^2((n-k)!)^2}$$

$$= \frac{(2n)!}{4^{2n}(n!)^2} \sum_{k=0}^{n} \binom{n}{k}^2$$

$$= \frac{(2n)!}{4^{2n}(n!)^2} \binom{2n}{n}$$

$$= \frac{((2n)!)^2}{4^{2n}(n!)^4} \simeq_{n\to\infty} \frac{1}{\pi n},$$

where we applied Stirling's approximation $n! \simeq (n/e)^n \sqrt{2\pi n}$ as n tends to infinity, and the combinatorial identity[2]

$$\binom{2n}{n} = \sum_{k=0}^{n} \binom{n}{k}^2, \qquad n \geqslant 0.$$

This yields

$$\sum_{n \geqslant 0} [P^{2n}]_{\vec{0}, \vec{0}} = \infty,$$

and we conclude by Corollary 1.12 which shows that $\mathbb{P}(T_{\vec{0}}^r < \infty) = 1$, see also Corollary 4.12 below.   □

Recurrence of d-Dimensional Random Walks, $d \geqslant 3$

We will use the following result, see Lemma 4 in Champion et al. (2007).

Lemma 4.7 *Let $n = a_n d + b_n$ where a_n is a nonnegative integer and $b_n \in \{0, 1, \ldots, d-1\}$. We have*

$$i_1! i_2! \cdots i_d! \geqslant (a_n!)^d (a_n + 1)^{b_n}$$

for all $i_1, i_2, \ldots, i_d$ nonnegative integers such that $i_1 + \cdots + i_d = n$, $d \geqslant 1$.

Proposition 4.8 *When $d \geqslant 3$, the symmetric random walk $(S_n)_{n \geqslant 0}$ is not recurrent, i.e. we have $\mathbb{P}(T_{\vec{0}}^r < \infty) < 1$.*

Proof By (4.3), we have

$$[P^{2n}]_{\vec{0}, \vec{0}} = \mathbb{P}(S_{2n} = \vec{0}) = \frac{1}{(2d)^{2n}} \sum_{\substack{i_1 + \cdots + i_d = n \\ i_1, i_2, \ldots, i_d \geqslant 0}} \frac{(2n)!}{(i_1!)^2 \cdots (i_d!)^2}.$$

Using the bound

$$i_1! i_2! \cdots i_d! \geqslant (a_n!)^d (a_n + 1)^{b_n}$$

---

[2] This identity can be proved by noting that the number $\binom{2n}{n}$ of ways to draw n balls among $2n$ balls can be obtained by summing the number of ways to draw exactly k white balls among n and $n - k$ black balls among n for $k = 0, 1, \ldots, n$.

for $n = i_1 + \cdots + i_d$ from Lemma 4.7 and the Euclidean division $n = a_n d + b_n$ where $b_n \in \{0, 1, \ldots, d-1\}$, we have

$$
\sum_{n \geqslant 1} [P^{2n}]_{\vec{0}, \vec{0}} = \sum_{n \geqslant 1} \frac{1}{(2d)^{2n}} \binom{2n}{n} \sum_{\substack{i_1 + \cdots + i_d = n \\ i_1, i_2, \ldots, i_d \geqslant 0}} \frac{(n!)^2}{(i_1!)^2 \cdots (i_d!)^2}
$$

$$
\leqslant \sum_{n \geqslant 1} \frac{1}{(2d)^{2n}} \binom{2n}{n} \frac{n!}{(a_n!)^d (a_n+1)^{b_n}} \sum_{\substack{i_1 + \cdots + i_d = n \\ i_1, i_2, \ldots, i_d \geqslant 0}} \frac{n!}{i_1! \cdots i_d!}
$$

$$
\leqslant \sum_{n \geqslant 1} \frac{1}{(2d)^{2n}} \binom{2n}{n} \frac{n! \, d^n}{(a_n!)^d a_n^{b_n}}
$$

$$
= \sum_{n \geqslant 1} \frac{(2n)!}{2^{2n} d^n n! (a_n!)^d a_n^{b_n}},
$$

from the formula

$$
d^n = \sum_{\substack{i_1 + \cdots + i_d = n \\ i_1, i_2, \ldots, i_d \geqslant 0}} \frac{n!}{i_1! \cdots i_d!}
$$

which follows from the multinomial identity

$$
\left(\sum_{l=1}^n x_l \right)^k = k! \sum_{\substack{d_1 + \cdots + d_n = k \\ d_1 \geqslant 0, \ldots, d_n \geqslant 0}} \frac{x_1^{d_1}}{d_1!} \cdots \frac{x_n^{d_n}}{d_n!}. \tag{4.13}
$$

Next, applying Stirling's approximation to $n!$, $(2n)!$ and $a_n!$, and using the limit $\lim_{m \to \infty} (1 + x/m)^m = e^x$, $x \in \mathbb{R}$, we have

$$
\frac{(2n)!}{2^{2n} d^n n! (a_n!)^d a_n^{b_n}} \simeq \frac{(2n/e)^{2n} \sqrt{4\pi n}}{2^{2n} d^n (n/e)^n \sqrt{2\pi n} ((a_n/e)^{a_n} \sqrt{2\pi a_n})^d a_n^{b_n}}
$$

$$
= \frac{\sqrt{2}}{(2\pi)^{d/2}} \frac{n^n d^{-n}}{e^{b_n} a_n^n a_n^{d/2}}
$$

$$
= \frac{\sqrt{2}}{(2\pi)^{d/2}} \frac{(1 - b_n/n)^{-n}}{e^{b_n} a_n^{d/2}}
$$

$$
\leqslant \frac{\sqrt{2} d^{d/2}}{(2\pi)^{d/2}} \frac{(1 - (d-1)/n)^{-n}}{(a_n d)^{d/2}}
$$

$$
\simeq \frac{\sqrt{2} d^{d/2} e^{d-1}}{(2\pi)^{d/2}} \frac{1}{n^{d/2}},
$$

since $a_n d \simeq n$ as n goes to infinity from the relation $a_n d/n = 1 - b_n/n$. We conclude that there exists a constant $C > 0$ such that for all n sufficiently large, we have

$$\frac{(2n)!}{2^{2n}d^n n!(a_n!)^d} \leqslant \frac{C}{n^{d/2}}, \tag{4.14}$$

hence the random walk is *not* recurrent when $d \geqslant 3$. Indeed, (4.14) shows that

$$\sum_{n \geqslant 0} \mathbb{P}\big(S_n = \vec{0}\big) < \infty,$$

hence $\mathbb{P}\big(T_{\vec{0}}^r = \infty\big) > 0$ by Corollary 1.12 or see also Corollary 4.12 below.   □

Recurrence Revisited

In Corollary 4.12 below we provide an alternative proof of Corollary 1.12.

Proposition 4.9 *The probability distribution* $\mathbb{P}\big(T_{\vec{0}}^r = n\big)$, $n \geqslant 1$, *satisfies the convolution equation*

$$\mathbb{P}\big(S_n = \vec{0}\big) = \sum_{k=2}^{n} \mathbb{P}\big(T_{\vec{0}}^r = k\big)\mathbb{P}\big(S_{n-k} = \vec{0}\big), \qquad n \geqslant 1.$$

Proof We partition the event $\{S_n = \vec{0}\}$ into

$$\{S_n = \vec{0}\} = \bigcup_{k=2}^{n} \{S_{n-k} = \vec{0},\ S_{n-k+1} \neq \vec{0}, \ldots, S_{n-1} \neq \vec{0},\ S_n = \vec{0}\}, \qquad n \geqslant 1,$$

according to the time of *last* return to state $\vec{0}$ *before* time n, with $\mathbb{P}\big(\{S_1 = \vec{0}\}\big) = 0$ since we are starting from $S_0 = \vec{0}$, see Fig. 4.7.

Fig. 4.7 Last return to state 0 at time $k = 10$

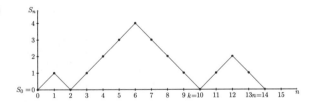

Then we have

$$\mathbb{P}(S_n = \vec{0}) := \mathbb{P}(S_n = \vec{0} \mid S_0 = \vec{0})$$

$$= \sum_{k=2}^{n} \mathbb{P}(S_{n-k} = \vec{0}, \ S_{n-k+1} \neq \vec{0}, \dots, S_{n-1} \neq \vec{0}, \ S_n = \vec{0} \mid S_0 = \vec{0})$$

$$= \sum_{k=2}^{n} \mathbb{P}(S_{n-k+1} \neq \vec{0}, \dots, S_{n-1} \neq \vec{0}, \ S_n = \vec{0} \mid S_{n-k} = \vec{0}, \ S_0 = \vec{0})$$

$$\times \mathbb{P}(S_{n-k} = \vec{0} \mid S_0 = \vec{0})$$

$$= \sum_{k=2}^{n} \mathbb{P}(S_1 \neq \vec{0}, \dots, S_{k-1} \neq \vec{0}, \ S_k = \vec{0} \mid S_0 = \vec{0}) \mathbb{P}(S_{n-k} = \vec{0} \mid S_0 = \vec{0})$$

$$= \sum_{k=2}^{n} \mathbb{P}(T_{\vec{0}}^r = k \mid S_0 = \vec{0}) \mathbb{P}(S_{n-k} = \vec{0} \mid S_0 = \vec{0})$$

$$= \sum_{k=2}^{n} \mathbb{P}(S_{n-k} = \vec{0}) \mathbb{P}(T_{\vec{0}}^r = k), \qquad n \geqslant 1.$$

$\square$

Lemma 4.10 *For all $m \geqslant 1$ we have*

$$1 - \frac{1}{\displaystyle\sum_{n=0}^{m} \mathbb{P}(S_n = \vec{0})} \leqslant \sum_{n=2}^{m} \mathbb{P}(T_{\vec{0}}^r = n) \leqslant \frac{\displaystyle\sum_{n=2}^{2m} \mathbb{P}(S_n = \vec{0})}{\displaystyle\sum_{n=0}^{m} \mathbb{P}(S_n = \vec{0})}. \qquad (4.15)$$

Proof We start by showing that

$$\sum_{n=1}^{m} \mathbb{P}(S_n = \vec{0}) = \sum_{k=2}^{m} \mathbb{P}(T_{\vec{0}}^r = k) \sum_{l=0}^{m-k} \mathbb{P}(S_l = \vec{0}).$$

We have

$$\sum_{n=1}^{m} \mathbb{P}(S_n = \vec{0}) = \sum_{n=1}^{m} \sum_{k=2}^{n} \mathbb{P}(T_{\vec{0}}^r = k) \mathbb{P}(S_{n-k} = \vec{0})$$

$$= \sum_{k=2}^{m} \sum_{n=k}^{m} \mathbb{P}(T_{\vec{0}}^r = k) \mathbb{P}(S_{n-k} = \vec{0})$$

$$= \sum_{k=2}^{m} \mathbb{P}(T_{\vec{0}}^r = k) \sum_{l=0}^{m-k} \mathbb{P}(S_l = \vec{0})$$

$$\leq \sum_{k=2}^{m} \mathbb{P}(T_{\vec{0}}^r = k) \sum_{l=0}^{m} \mathbb{P}(S_l = \vec{0})$$

$$= \left(\sum_{n=0}^{m} \mathbb{P}(S_n = \vec{0}) \right) \left(\sum_{n=2}^{m} \mathbb{P}(T_{\vec{0}}^r = n) \right).$$

On the other hand, we have

$$\sum_{n=1}^{2m} \mathbb{P}(S_n = \vec{0}) = \sum_{n=2}^{2m} \mathbb{P}(T_{\vec{0}}^r = n) \sum_{l=0}^{2m-n} \mathbb{P}(S_l = \vec{0})$$

$$\geq \sum_{n=2}^{m} \mathbb{P}(T_{\vec{0}}^r = n) \sum_{l=0}^{2m-n} \mathbb{P}(S_l = \vec{0})$$

$$\geq \sum_{n=2}^{m} \mathbb{P}(T_{\vec{0}}^r = n) \sum_{l=0}^{m} \mathbb{P}(S_l = \vec{0}).$$

□

By letting m tend to ∞ in (4.15) we get the following corollary.

Corollary 4.11 *We have*

$$\mathbb{P}(T_{\vec{0}}^r < \infty) = 1 - \frac{1}{\sum_{n \geq 0} \mathbb{P}(S_n = \vec{0})} = 1 - \frac{1}{1 + \mathbb{E}[R_{\vec{0}} \mid S_0 = \vec{0}]}.$$

Proof By Lemma 4.10, letting m tend to infinity in (4.15), we have

$$1 - \frac{1}{\sum_{n \geq 0} \mathbb{P}(S_n = \vec{0})} \leq \sum_{n \geq 2} \mathbb{P}(T_{\vec{0}}^r = n)$$

$$= \mathbb{P}(T_{\vec{0}}^r < \infty)$$

$$\leq \frac{\sum_{n \geq 2} \mathbb{P}(S_n = \vec{0})}{\sum_{n \geq 0} \mathbb{P}(S_n = \vec{0})}$$

$$= 1 - \frac{1}{\sum\limits_{n \geqslant 0} \mathbb{P}(S_n = \vec{0})}.$$

□

The following result is a consequence of Corollary 4.11. Note that the sum of the series

$$\sum_{n \geqslant 0} \mathbb{P}(S_n = \vec{0}) = \sum_{n \geqslant 0} \mathbb{E}[\mathbb{1}_{\{S_n = \vec{0}\}}] = \mathbb{E}\left[\sum_{n \geqslant 0} \mathbb{1}_{\{S_n = \vec{0}\}}\right]$$

represents the average number of visits to state $\textcircled{0}$, see also Corollary 1.12. We also have

$$\sum_{n \geqslant 0} \mathbb{P}(S_n = \vec{0}) = \sum_{n \geqslant 0} [P^n]_{0,0} = (I - P)_{0,0}^{-1}.$$

Corollary 4.12 *The d-dimensional symmetric random walk is recurrent, i.e.* $\mathbb{P}(T_{\vec{0}}^r < \infty) = 1$, *if and only if*

$$\sum_{n \geqslant 0} \mathbb{P}(S_n = \vec{0}) = \infty.$$

4.3   Reflected Random Walk

We now consider a reflected random walk $(S_n)_{n \geqslant 0}$ with transition probabilities

$$\begin{cases} \mathbb{P}(S_{n+1} = k + 1 \mid S_n = k) = p, & k = 0, 1, \ldots, L - 1, \\[2mm] \mathbb{P}(S_{n+1} = k - 1 \mid S_n = k) = q, & k = 1, 2, \ldots, L - 1, \end{cases}$$

with

$$\mathbb{P}(S_{n+1} = 0 \mid S_n = 0) = q \quad \text{and} \quad \mathbb{P}(S_{n+1} = L \mid S_n = L) = 1,$$

for all $n \in \mathbb{N} = \{0, 1, 2, \ldots\}$, where $q = 1 - p$ and $p \in (0, 1]$.

Proposition 4.13 *If $p \in (0, 1]$, state $\textcircled{L}$ is eventually reached in finite time with probability one after starting from any state $k \in \{0, 1, \ldots, L\}$.*

Proof Let

$$g(k) := \mathbb{P}(T_L < \infty \mid S_0 = k)$$

denote the probability that state (L) is reached in finite time after starting from state $k \in \{0, 1, \ldots, L\}$. Using first step analysis we can write down the difference equations satisfied by $g(k)$, $k = 0, 1, \ldots, L - 1$, as

$$\begin{cases} g(k) = pg(k + 1) + qg(k - 1), & k = 1, 2, \ldots, L - 1, & (4.16\text{a}) \\ \\ g(0) = pg(1) + qg(0), & & (4.16\text{b}) \end{cases}$$

with the boundary condition $g(L) = 1$. In order to solve for the solution $g(k) := \mathbb{P}(T_L < \infty \mid S_0 = k)$ of (4.16a)–(4.16b), $k = 0, 1, \ldots, L$, we observe that the constant function $g(k) = C$ is solution of both (4.16a) and (4.16b) and the boundary condition $g(L) = 1$ yields $C = 1$, hence

$$g(k) = \mathbb{P}(T_L < \infty \mid S_0 = k) = 1$$

for all $k = 0, 1, \ldots, L$. $\square$

Let

$$h(k) := \mathbb{E}[T_L \mid S_0 = k]$$

denote the expected time until state (L) is reached after starting from state $k \in \{0, 1, \ldots, L\}$.

Proposition 4.14 *We have*

$$h(k) = \mathbb{E}[T_L \mid S_0 = k] = \frac{L - k}{p - q} + \frac{q}{(p - q)^2} \left(\left(\frac{q}{p} \right)^L - \left(\frac{q}{p} \right)^k \right),$$

$k = 0, 1, \ldots, L$, *when* $p \neq q$, *and*

$$h(k) = \mathbb{E}[T_L \mid S_0 = k] = (L + k + 1)(L - k), \quad k = 0, 1, \ldots, L,$$

when $p = q = 1/2$.

Proof Using first step analysis we can write down the difference equations satisfied by $h(k)$ for $k = 0, 1, \ldots, L - 1$, as

$$\begin{cases} h(k) = 1 + ph(k + 1) + qh(k - 1), & k = 1, 2, \ldots, L - 1, & (4.17\text{a}) \\[2em] h(0) = 1 + ph(1) + qh(0), & & (4.17\text{b}) \end{cases}$$

with the boundary condition $h(L) = 0$. We compute $h(k) = \mathbb{E}[T_L \mid S_0 = k]$ for all $k = 0, 1, \ldots, L$ by solving the Eqs. (4.17a)–(4.17b) for $k = 1, 2, \ldots, L - 1$.

(i) Case $p \neq q$. The solution of the associated *homogeneous equation*

$$h(k) = ph(k + 1) + qh(k - 1), \qquad k = 1, 2, \ldots, L - 1, \qquad (4.18)$$

has the form

$$h(k) = C_1 + C_2(q/p)^k, \qquad k = 1, 2, \ldots, L - 1.$$

In addition, we can check that $k \mapsto k/(p-q)$ is a particular solution of (4.17a). Hence the general solution of (4.17a) is written as the sum

$$h(k) = \frac{k}{q - p} + C_1 + C_2(q/p)^k, \qquad k = 0, 1, \ldots, L,$$

which can be obtained in Mathematica via the command

$$\text{RSolve[f[k]=1+pf[k+1]+(1-p)f[k-1],f[k],k],}$$

with

$$\begin{cases} 0 = h(L) = \dfrac{L}{q - p} + C_1 + C_2(q/p)^L, \\[2em] ph(0) = p(C_1 + C_2) = 1 + ph(1) = 1 + p\left(\dfrac{1}{q - p} + C_1 + C_2\dfrac{q}{p}\right), \end{cases}$$

which yields

$$\begin{cases} C_1 = q\dfrac{(q/p)^L}{(p - q)^2} - \dfrac{L}{q - p}, \\[2em] C_2 = -\dfrac{q}{(p - q)^2}, \end{cases}$$

and

$$h(k) = \mathbb{E}[T_L \mid S_0 = k] = \frac{L-k}{p-q} + \frac{q}{(p-q)^2}((q/p)^L - (q/p)^k),$$

$k = 0, 1, \ldots, L$.

(ii) Case $p = q = 1/2$. The solution of the associated *homogeneous equation* (4.18) is given by

$$h(k) = C_1 + C_2 k, \qquad k = 1, 2, \ldots, L-1,$$

and the general solution to (4.17a) has the form

$$h(k) = -k^2 + C_1 + C_2 k, \qquad k = 1, 2, \ldots, L,$$

which can be obtained in Mathematica via the command

$$\text{RSolve[g[k]=1+(1/2)g[k+1]+(1/2)g[k-1],g[k],k],}$$

with

$$\begin{cases} 0 = h(L) = -L^2 + C_1 + C_2 L, \\[2mm] \dfrac{h(0)}{2} = \dfrac{C_1}{2} = 1 + \dfrac{h(1)}{2} = 1 + \dfrac{-1 + C_1 + C_2}{2}, \end{cases}$$

hence

$$\begin{cases} C_1 = L(L+1), \\[2mm] C_2 = -1, \end{cases}$$

which yields

$$h(k) = \mathbb{E}[T_L \mid S_0 = k] = (L+k+1)(L-k), \quad k = 0, 1, \ldots, L.$$

$\square$

As a consequence of Proposition 5.2 below, the reflected random walk is recurrent when $p \leqslant 1/2$, and transient when $p > 1/2$.
Letting $\varepsilon = 1 - q/p$, i.e. $q/p = 1 + \varepsilon$, we check that, as ε tends to zero,

$$\frac{L-k}{p-q} + \frac{q}{(p-q)^2}((q/p)^L - (q/p)^k)$$

$$= -\frac{L-k}{\varepsilon p} - (1+\varepsilon)^{k+1}\frac{1}{p\varepsilon^2}(1 - (1+\varepsilon)^{L-k})$$

$$= -\frac{L-k}{\varepsilon p} - (1 + (k+1)\varepsilon)\frac{1}{p\varepsilon^2}(-(L-k)\varepsilon - (L-k)(L-k-1)\varepsilon^2/2)$$

$$= \frac{1}{p}((L-k)(L-k-1)/2) + (k+1)\frac{1}{p}(L-k)$$

$$\simeq (L-k)(L-k-1) + 2(k+1)(L-k)$$

$$= (L-k)(L+k+1).$$

4.4 Conditioned Random Walk

Conditional Hitting Probabilities

Consider the one-dimensional random walk $(S_n)_{n \geqslant 0}$, let

$$T_L := \inf\{n \geqslant 0 \ : \ S_n = L\}$$

denote the first hitting time of L by the process $(S_n)_{n \geqslant 0}$, and let

$$T_0 := \inf\{n \geqslant 0 \ : \ S_n = 0\}$$

denote the first hitting time of 0 by the process $(S_n)_{n \geqslant 0}$ (Fig. 4.8).

Lemma 4.15 *The probability of an upward step from state* $\textcircled{k}$ *given that state L is reached first, is given by*

$$\mathbb{P}(S_1 = k + 1 \mid S_0 = k \text{ and } T_L < T_0) = p + \frac{p-q}{(p/q)^k - 1},$$

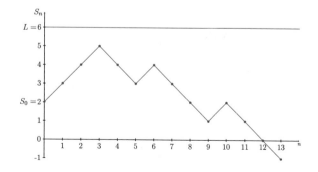

Fig. 4.8 Sample path of the random walk $(S_n)_{n \geqslant 0}$

when $p \neq q$, and by

$$\mathbb{P}(S_1 = k+1 \mid S_0 = k \text{ and } T_L < T_0) = \frac{1}{2} + \frac{1}{2k},$$

when $p = q = 1/2$, $k = 1, 2, \ldots, L-1$.

Proof We note the equality

$$\mathbb{P}(T_L < T_0 \mid S_1 = k+1 \text{ and } S_0 = k) = \mathbb{P}(T_L < T_0 \mid S_1 = k+1)$$
$$= \mathbb{P}(T_L < T_0 \mid S_0 = k+1). \qquad (4.19)$$

for $k \in \{0, 1, \ldots, L-1\}$. Indeed, given that we start from state $\boxed{k+1}$ at time 1, whether $T_L < T_0$ or $T_L > T_0$ does not depend on the past of the process before time 1. In addition, it does not matter whether we start from state $\boxed{k+1}$ at time 1 or at time 0. Hence, we have

$$\mathbb{P}(S_1 = k+1 \mid S_0 = k \text{ and } T_L < T_0) = \frac{\mathbb{P}(S_1 = k+1, \ S_0 = k, \ T_L < T_0)}{\mathbb{P}(S_0 = k \text{ and } T_L < T_0)}$$

$$= \frac{\mathbb{P}(T_L < T_0 \mid S_1 = k+1 \text{ and } S_0 = k)\mathbb{P}(S_1 = k+1 \text{ and } S_0 = k)}{\mathbb{P}(T_L < T_0 \text{ and } S_0 = k)}$$

$$= p\frac{\mathbb{P}(T_L < T_0 \mid S_1 = k+1 \text{ and } S_0 = k)}{\mathbb{P}(T_L < T_0 \mid S_0 = k)}$$

$$= p\frac{\mathbb{P}(T_L < T_0 \mid S_0 = k+1)}{\mathbb{P}(T_L < T_0 \mid S_0 = k)}$$

$$= p\frac{p_{k+1}}{p_k}, \qquad k = 0, 1, 2, \ldots, L-1, \qquad (4.20)$$

where

$$p_k := \mathbb{P}(T_L < T_0 \mid S_0 = k), \qquad k = 0, 1, \ldots, L.$$

We conclude by the relations

$$p_k := \mathbb{P}(T_L < T_0 \mid S_0 = k) = \frac{1 - (q/p)^k}{1 - (q/p)^L}, \qquad k = 0, 1, \ldots, L, \qquad (4.21)$$

when $p \neq q$, see Proposition 1.24, and by

$$p_k = \frac{k}{L}, \qquad k = 0, 1, \ldots, L, \qquad (4.22)$$

when $p = q = 1/2$, see Relation (1.44). $\qquad \square$

By exchanging states $\textcircled{0}$ and $\textcircled{L}$ we also obtain the following result.

Lemma 4.16 *The probability of a downward step from state $\textcircled{k}$ given that state $\textcircled{0}$ is reached first is given by*

$$\mathbb{P}(S_1 = k - 1 \mid S_0 = k \text{ and } T_0 < T_L) = q + \frac{q - p}{(q/p)^{L-k} - 1},$$

when $p \neq q$, and by

$$\mathbb{P}(S_1 = k - 1 \mid S_0 = k \text{ and } T_0 < T_L) = \frac{1}{2} + \frac{1}{2(L - k)},$$

when $p = q = 1/2$, $k = 1, 2, \ldots, L - 1$.

Proof We compute the probability

$$\mathbb{P}(S_1 = k - 1 \mid S_0 = k \text{ and } T_0 < T_L), \qquad k = 1, 2, \ldots, L,$$

of a downward step given that state $\textcircled{0}$ is reached first. We have

$$
\begin{aligned}
&\mathbb{P}(S_1 = k - 1 \mid S_0 = k \text{ and } T_0 < T_L) \\
&= \frac{\mathbb{P}(S_1 = k - 1, \ S_0 = k \text{ and } T_0 < T_L)}{\mathbb{P}(S_0 = k \text{ and } T_0 < T_L)} \\
&= \frac{\mathbb{P}(T_0 < T_L \mid S_1 = k - 1 \text{ and } S_0 = k)\mathbb{P}(S_1 = k - 1 \text{ and } S_0 = k)}{\mathbb{P}(T_0 < T_L \text{ and } S_0 = k)} \\
&= q\frac{\mathbb{P}(T_0 < T_L \mid S_0 = k - 1)}{\mathbb{P}(T_0 < T_L \mid S_0 = k)} \\
&= q\frac{1 - p_{k-1}}{1 - p_k}, \qquad k = 1, 2, \ldots, L - 1,
\end{aligned}
$$

and we conclude using (4.21) and (4.22), see Propositions 1.24 and (1.44). $\square$

Similarly, we can compute the probability of a downward step from state $\textcircled{k}$ given that state $\textcircled{L}$ is reached first as

$$\mathbb{P}(S_1 = k - 1 \mid S_0 = k \text{ and } T_L < T_0) = 1 - \mathbb{P}(S_1 = k + 1 \mid S_0 = k \text{ and } T_L < T_0)$$

$$= q + \frac{q - p}{(p/q)^k - 1},$$

when $p \neq q$, and as

$$\mathbb{P}(S_1 = k - 1 \mid S_0 = k \text{ and } T_L < T_0) = 1 - \mathbb{P}(S_1 = k + 1 \mid S_0 = k \text{ and } T_L < T_0)$$

$$= \frac{1}{2} - \frac{1}{2k},$$

when $p = q = 1/2$, $k = 1, 2, \ldots, L - 1$.

Conditional Mean Hitting Times

Let now

$$T_L := \inf\{n \geqslant 0 \ : \ S_n = L\}$$

denote the first hitting time of state $\textcircled{L}$, with $T_L = +\infty$ in case state $\textcircled{L}$ is never reached, see Fig. 4.9.
Let

$$h(k) = \mathbb{E}[T_L \mid S_0 = k, \ T_L < T_0], \qquad k = 1, 2, \ldots, L,$$

denote the expected value of T_L given that state $\textcircled{0}$ is never reached. The next result will be used for the proof of Proposition 5.10 on cookie-excited random walks.

Proposition 4.17 *When* $p \neq q$, *we have*

$$h(k) = \mathbb{E}[T_L \mid S_0 = k, \ T_L < T_0]$$

$$= \frac{(1 - (q/p)^k)L(1 + (q/p)^L) - k(1 + (q/p)^k)(1 - (q/p)^L)}{(p - q)(1 - (q/p)^L)(1 - (q/p)^k)}$$

$$= \frac{L(1 + (q/p)^L)}{(p - q)(1 - (q/p)^L)} - \frac{k(1 + (q/p)^k)}{(p - q)(1 - (q/p)^k)},$$

Fig. 4.9 Sample paths of the random walk $(S_n)_{n \geqslant 0}$

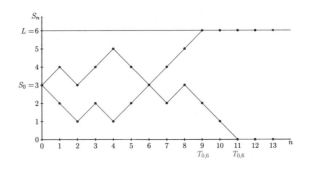

whereas when $p = q = 1/2$ we find

$$h(k) = \mathbb{E}[T_L \mid S_0 = k, \ T_L < T_0] = \frac{L^2 - k^2}{3}, \qquad k = 1, 2, \ldots, L.$$

Proof Using the transition probabilities (4.20) we state the finite difference equations satisfied by $h(k)$, $k = 1, 2, \ldots, L - 1$, as

$$h(k) = 1 + h(k + 1)\mathbb{P}(S_1 = k + 1 \mid S_0 = k \text{ and } T_L < T_0)$$
$$+ h(k - 1)\mathbb{P}(S_1 = k - 1 \mid S_0 = k \text{ and } T_L < T_0)$$
$$= 1 + p\frac{p_{k+1}}{p_k}h(k + 1) + \left(1 - p\frac{p_{k+1}}{p_k}\right)h(k - 1), \qquad (4.23)$$

$k = 1, 2, \ldots, L - 1$, or, due to the first step equation $p_k = pp_{k+1} + qp_{k-1}$,

$$p_k h(k) = p_k + pp_{k+1}h(k + 1) + qp_{k-1}h(k - 1), \qquad k = 1, 2, \ldots, L - 1,$$

with the boundary condition $h(L) = 0$. Letting $g(k) := p_k h(k)$, we check that $g(k)$ satisfies

$$g(k) = p_k + pg(k + 1) + qg(k - 1), \qquad k = 1, 2, \ldots, L - 1, \qquad (4.24)$$

with the boundary conditions $g(0) = 0$ and $g(L) = 0$.

(i) When $p = q = 1/2$ we have $p_k = k/L$ by (4.22), hence (4.23) becomes

$$h(k) = 1 + \frac{k + 1}{2k}h(k + 1) + \frac{k - 1}{2k}h(k - 1),$$

$k = 1, 2, \ldots, L - 1$, and (4.24) can be written as

$$g(k) = \frac{k}{L} + \frac{1}{2}g(k + 1) + \frac{1}{2}g(k - 1), \qquad k = 1, 2, \ldots, L - 1, \qquad (4.25)$$

with the boundary conditions $g(0) = 0$ and $g(L) = 0$. We check that $g(k) = Ck^3$ is a particular solution of (4.25) when $C = -1/(3L)$, hence the general solution of (4.25) takes the form

$$g(k) = -\frac{k^3}{3L} + C_1 + C_2 k,$$

where C_1 and C_2 are determined by the boundary conditions

$$0 = g(0) = C_1$$

and

$$0 = g(L) = -\frac{1}{3}L^2 + C_1 + C_2 L,$$

i.e. $C_1 = 0$ and $C_2 = L/3$. Consequently, we have

$$g(k) = \frac{k}{3L}(L^2 - k^2), \qquad k = 0, 1, \ldots, L,$$

hence we have

$$h(k) = \mathbb{E}[T_L \mid S_0 = k, \ T_L < T_0] = \frac{L^2 - k^2}{3}, \qquad k = 1, 2, \ldots, L,$$

which can be obtained in Mathematica via the command

RSolve[g[k]=k/L+(1/2)g[k+1]+(1/2)g[k-1],g[k],k].

(ii) When $p \neq q$, by (4.21) we have

$$p_k = \frac{1 - (q/p)^k}{1 - (q/p)^L}, \qquad k = 0, 1, \ldots, L,$$

hence (4.23) can be rewritten as

$$h(k) = 1 + p\frac{1 - (q/p)^{k+1}}{1 - (q/p)^k}h(k+1) + q\frac{1 - (q/p)^{k-1}}{1 - (q/p)^k}h(k-1),$$

and (4.24) can be rewritten as

$$g(k) = \frac{1 - (q/p)^k}{1 - (q/p)^L} + pg(k+1) + qg(k-1), \tag{4.26}$$

$k = 1, 2, \ldots, L - 1$, with

$$g(k) = (1 - (q/p)^k)h(k).$$

We check that

$$g(k) := -\frac{(p-q)k(1 + (q/p)^k) + p - q(q/p)^k}{(p-q)^2(1 - (q/p)^L)},$$

$k = 0, 1, \ldots, L$, is a particular solution of (4.26), hence the general solution of (4.26) takes the form

$$g(k) = -\frac{(p-q)k(1+(q/p)^k) + p - q(q/p)^k}{(p-q)^2(1-(q/p)^L)} + C_1 + C_2(q/p)^k,$$

$k = 0, 1, \ldots, L$, under the boundary conditions

$$g(0) = 0 = -\frac{1}{p-q} + C_1 + C_2$$

and

$$g(L) = 0$$

$$= -\frac{(p-q)L(1+(q/p)^L) + p - q(q/p)^L}{(p-q)^2(1-(q/p)^L)} + C_1 + C_2(q/p)^L$$

$$= -\frac{(p-q)L(1+(q/p)^L) + p - q(q/p)^L}{(p-q)^2(1-(q/p)^L)}$$

$$+ \frac{1}{p-q} - C_2(1-(q/p)^L)$$

$$= -\frac{(p-q)L(1+(q/p)^L) + q - q(q/p)^L}{(p-q)^2(1-(q/p)^L)} - C_2(1-(q/p)^L),$$

or

$$C_1 = \frac{(p-q)L(1+(q/p)^L) + (1-(q/p)^L)p}{(p-q)^2(1-(q/p)^L)^2},$$

and

$$C_2 = -\frac{(p-q)L(1+(q/p)^L) + q(1-(q/p)^L)}{(p-q)^2(1-(q/p)^L)^2},$$

hence

$$g(k) = -\frac{(p-q)k(1+(q/p)^k) + p - q(q/p)^k}{(p-q)^2(1-(q/p)^L)}$$

$$+ \frac{(p-q)L(1+(q/p)^L) + (1-(q/p)^L)p}{(p-q)^2(1-(q/p)^L)^2}$$

$$- (q/p)^k \frac{(p-q)L(1+(q/p)^L) + (1-(q/p)^L)q}{(p-q)^2(1-(q/p)^L)^2}$$

$$= -\frac{(p-q)k(1+(q/p)^k)+p-q(q/p)^k}{(p-q)^2(1-(q/p)^L)}$$

$$+(1-(q/p)^k)\frac{(p-q)L(1+(q/p)^L)}{(p-q)^2(1-(q/p)^L)^2}$$

$$+(1-(q/p)^L)\frac{p-q(q/p)^k}{(p-q)^2(1-(q/p)^L)^2}$$

$$= \frac{(1-(q/p)^k)L(1+(q/p)^L)-k(1+(q/p)^k)(1-(q/p)^L)}{(p-q)(1-(q/p)^L)^2},$$

$k = 0, 1, \ldots, L$, and

$$h(k) = \frac{(1-(q/p)^k)L(1+(q/p)^L)-k(1+(q/p)^k)(1-(q/p)^L)}{(p-q)(1-(q/p)^L)(1-(q/p)^k)},$$

$k = 1, 2, \ldots, L$, which can be obtained in Mathematica via the command

RSolve[g[k]=1-(q/p)^k+(1/2)g[k+1]+(1/2)g[k-1],g[k],k].

$\qquad\qquad\qquad\qquad\qquad\qquad\qquad\qquad\qquad\qquad\qquad\qquad\qquad\qquad\square$

Letting $\varepsilon = 1 - q/p$, i.e. $q/p = 1 + \varepsilon$, we have, as ε tends to zero,

$$h(k) \simeq \frac{(1-(1+\varepsilon)^k)L(1+(1+\varepsilon)^L)-k(1+(1+\varepsilon)^k)(1-(1+\varepsilon)^L)}{(p-q)(1-(1+\varepsilon)^L)(1-(1+\varepsilon)^k)}$$

$$= \frac{(k\varepsilon+k(k-1)\varepsilon^2/2+k(k-1)(k-2)\varepsilon^3/6)L(2+L\varepsilon+L(L-1)\varepsilon^2/2)}{p\varepsilon^3}$$

$$- \frac{k(2+k\varepsilon+k(k-1)\varepsilon^2/2)(L\varepsilon+L(L-1)\varepsilon^2/2+L(L-1)(L-2)\varepsilon^3/6)}{p\varepsilon^3}$$

$$= \frac{L^2-k^2}{6p}$$

$$\simeq \frac{L^2-k^2}{3}, \quad k = 0, 1, \ldots, L.$$

The conditional expectation $h(0)$ is actually undefined because the event $\{S_0 = 0, T_L < T_0\}$ has probability 0.

Notes

See § 1.2 and Proposition 1.3 in Hairer (2016) for the general theory of recurrence of Markov chains and their application to random walks.

Exercises

Exercise 4.1 Consider a sequence $(X_n)_{n \geqslant 0}$ of independent $\{0, 1\}$-valued Bernoulli random variables with distribution $\mathbb{P}(X_n = 1) = p$, $\mathbb{P}(X_n = 0) = q$, $n \geqslant 1$.

(a) Show that

$$\mathbb{E}\left[\exp\left(t \sum_{k=1}^{n} X_k \right) \right] = (q + pe^t)^n, \quad n \geqslant 0, \quad t \in \mathbb{R}.$$

(b) Using the Markov inequality, show that

$$\mathbb{P}\left(\frac{1}{n} \sum_{k=1}^{n} (X_k - p) \geqslant z \right) \leqslant e^{-n((p+z)t - \log(q + pe^t))}, \qquad z > 0, \, t > 0.$$

(c) Find the value $t(x)$ of $t > 0$ that maximizes $t \mapsto xt - \log(q + pe^t)$ for x fixed in $(0, 1)$.

(d) Show the bound

$$\mathbb{P}\left(\frac{1}{n} \sum_{k=1}^{n} (X_k - p) \geqslant z \right) \leqslant \exp\left(-n \left((p + z) \log \frac{(p + z)q}{(q - z)p} - \log \frac{q}{q - z} \right) \right),$$

$0 \leqslant z < q$.

(e) Using Taylor's formula with remainder

$$f(t) = f(0) + tf'(0) + \frac{t^2}{2} f''(\theta t), \qquad t \in \mathbb{R},$$

for some $\theta \in [0, 1]$, show that $\log(q + pe^t) \leqslant pt + t^2/8$, $t \in \mathbb{R}$.
Hint. Show that for all $\alpha \in \mathbb{R}$ we have $4pq\alpha \leqslant (q + p\alpha)^2$.

(f) Find the value $t(z)$ of $t \in \mathbb{R}$ that maximizes $t \mapsto zt - t^2/8$ for $z \in \mathbb{R}$.

(g) Show the bound

$$\mathbb{P}\left(\frac{1}{n} \sum_{k=1}^{n} (X_k - p) \geqslant z \right) \leqslant e^{-2nz^2}, \qquad z \geqslant 0. \tag{4.27}$$

Problem 4.2 Multi-Armed Bandits (MABs) have applications from recommender systems and information retrieval to healthcare and finance, due to its stellar performance combined with attractive properties, such as learning from less feedback, see Bouneffouf et al. (2020). For example, the Uber Data Science team leverages MAB testing to rank restaurants on the main feed of the Uber Eats app.

We consider an N-arm bandit in which the reward of arm n°i at time $n \geq 1$ is $X_n^{(i)}$, where for $i = 1, \ldots, N$, $(X_n^{(i)})_{n \geq 0}$ is a i.i.d. Bernoulli sequence with $\mathbb{P}(X_n^{(i)} = 1) = p_i \in [0, 1]$, $n \geq 1$, ordered as $p_1 \leq \cdots \leq p_N$. We let

$$\widehat{m}_n^{(i,\alpha)} := \frac{1}{T_n^{(i,\alpha)}} \sum_{k=1}^n X_k^{(i)} \mathbb{1}_{\{\alpha_k = i\}}$$

denote the sample average reward obtained from arm n°i until time $n \geq 1$ under a given policy $(\alpha_k)_{k \geq 1}$. We define the policy $(\alpha_n^*)_{n \geq 1}$ by $\alpha_n^* := n$ for $n = 1, \ldots, N$, and for $n > N$ we let α_n^* be the index $i \in \{1, \ldots, N\}$ that maximizes the quantity $\widehat{m}_{n-1}^{(i,\alpha^*)} + \sqrt{2(\log n)/T_{n-1}^{(i,\alpha^*)}}$.

(a) Let $1 \leq i < N$ and $n \geq N$. Show by contradiction that if $\alpha_n^* = i$, then at least one of the three following conditions must hold:

$$\widehat{m}_{n-1}^{(N,\alpha^*)} + \sqrt{\frac{2 \log n}{T_{n-1}^{(N,\alpha^*)}}} \leq p_N, \quad \widehat{m}_{n-1}^{(i,\alpha^*)} > p_i + \sqrt{\frac{2 \log n}{T_{n-1}^{(i,\alpha^*)}}}, \quad T_{n-1}^{(i,\alpha^*)} < \frac{8 \log n}{(p_N - p_i)^2}.$$

(b) Show that letting $\widehat{n}_i := \lceil 8(\log n)/(p_N - p_i)^2 \rceil$, we have

$$\mathbb{E}\big[T_n^{(i,\alpha^*)}\big] \leq \widehat{n}_i + \sum_{\widehat{n}_i < k \leq n} \left(\mathbb{P}\left(\widehat{m}_{k-1}^{(N,\alpha^*)} + \sqrt{\frac{2 \log k}{T_{k-1}^{(N,\alpha^*)}}} \leq p_N \right) \right.$$

$$\left. + \mathbb{P}\left(\widehat{m}_{k-1}^{(i,\alpha^*)} > p_i + \sqrt{\frac{2 \log k}{T_{k-1}^{(i,\alpha^*)}}} \right) \right), \quad 1 \leq 1 < N, \quad n \geq N.$$

(c) Show that $\mathbb{P}\left(\widehat{m}_{k-1}^{(N,\alpha^*)} + \sqrt{\frac{2 \log k}{T_{k-1}^{(N,\alpha^*)}}} \leq p_N \right) \leq \frac{1}{k^3}$ and

$$\mathbb{P}\left(\widehat{m}_{k-1}^{(i,\alpha^*)} > p_i + \sqrt{\frac{2 \log k}{T_{k-1}^{(i,\alpha^*)}}} \right) \leq \frac{1}{k^3}, \quad i = 1, \ldots, N, \quad k \geq N.$$

Hint. Use the bound (4.27) in Exercise 4.1

(d) Show that the *modified* regret, defined as

$$\overline{\mathcal{R}}_n^\alpha := \sum_{k=1}^n \mathbb{E}[p_N - p_{\alpha_k}],$$

can be bounded by

$$\overline{\mathcal{R}}_n^{\alpha^*} \leqslant \sum_{i=1}^{N-1} (p_N - p_i) + 8 \sum_{i=1}^{N-1} \frac{\log n}{p_N - p_i}, \qquad n \geqslant 1.$$

Hint. Use a comparison argument between series and integrals.

Problem 4.3

(a) Consider a gambling process $(S_n)_{n\geqslant0}$ taking values in the discrete interval $\{0, 1, \dots, L\}$ with respective probabilities p, q of increment and decrement. We let $T_{0,L}$ denote the hitting time of the boundary $\{0, L\}$ by $(S_n)_{n\geqslant0}$.

  (i) Compute the probability generating function

$$G_i(s) := \mathbb{E}\big[s^{T_{0,L}} \,\big|\, S_0 = i\big], \quad i = 0, 1, \dots, L, \quad s \in [-1, 1],$$

  of $T_{0,L}$. Consider the cases $p = q$ and $p \neq q$ separately.
  Hint. See Exercise 3.4 in Privault (2018).

  (ii) Compute the Laplace transform

$$L_i(\lambda) := \mathbb{E}\big[e^{-\lambda T_{0,L}} \,\big|\, S_0 = i\big], \qquad i = 0, 1, \dots, L, \quad \lambda \geqslant 0.$$

  of $T_{0,L}$. Consider the cases $p = q$ and $p \neq q$ separately.

(b) We rescale the process $(S_n)_{n\geqslant1}$ into a continuous-time random walk $(X_t)_{t\in\mathbb{R}_+}$. For this,

  • we split the time interval $[0, t]$ into $n \simeq t/\varepsilon$ time steps of length $\varepsilon > 0$,
  • we split the space interval $[0, y]$ into $L \simeq y/\sqrt{\varepsilon}$ steps of height $\sqrt{\varepsilon}$,
  • we rescale the probabilities p and q as

$$p_\varepsilon := \frac{1}{2}(1 - \mu\sqrt{\varepsilon}) \quad \text{and} \quad q_\varepsilon := \frac{1}{2}(1 + \mu\sqrt{\varepsilon}),$$

  for some $\mu \in \mathbb{R}$, see Equation (7.7) in Privault (2022), and we let ε tend to zero. We let $T_{0,y}$ denote the hitting time of the boundary $\{0, y\}$ by $(X_t)_{t\in\mathbb{R}_+}$.

  (i) Taking $\mu = 0$, compute the Laplace transform

$$L_x(\lambda) := \mathbb{E}\big[e^{-\lambda T_{0,y}} \,\big|\, X_0 = x\big], \qquad x \in [0, y], \quad \lambda \geqslant 0,$$

  of $T_{0,y}$.
  Hint. The answer should recover Equation (3) in Antal and Redner (2005), see also Equation (2.2.10) in Redner (2001) and Exercise 14.3-a) in Privault (2022).

(ii) Compute the Laplace transform

$$L_x(\lambda) := \mathbb{E}\big[e^{-\lambda T_{0,y}} \,\big|\, X_0 = x\big], \qquad x \in [0, y], \quad \lambda \geqslant 0,$$

of $T_{0,y}$ in case $\mu \neq 0$.

Hint. See also Exercise 14.5 in Privault (2022).

(c) Repeat Questions (a) and (b) above *for the hitting time T_L of the level L* when $(S_n)_{n \geqslant 0}$ is the random walk on $\{0, 1, \ldots, L\}$ reflected at state 0.

Hint. Your answer should recover Equation (5) in Antal and Redner (2005) when $\mu = 0$, see also Equation (2.2.21) in Redner (2001).

Problem 4.4 Consider a random walk $(S_n)_{n \geqslant 0}$ on $\mathbb{Z}$ with independent increments, such that

$$\mathbb{P}(S_{n+1} - S_n = +1) = p \quad \text{and} \quad \mathbb{P}(S_{n+1} - S_n = -1) = q, \qquad n \geqslant 0,$$

with $p + q = 1$. The sequence $(T_0^k)_{k \geqslant 1}$ of return times to 0 of $(S_n)_{n \geqslant 0}$ is defined recursively with

$$T_0^1 := \inf\{n \geqslant 1 \,:\, S_n = 0\}.$$

and

$$T_0^{k+1} := \inf\{n > T_0^k \,:\, S_n = 0\}, \qquad k \geqslant 1.$$

(a) Consider the generating function $H_i(s)$ defined as

$$H_i(s) := \mathbb{E}\left[\sum_{k \geqslant 1} s^{T_0^k} \,\middle|\, S_0 = i\right], \qquad i \in \mathbb{Z}, \quad -1 \leqslant s \leqslant 1.$$

Using first step analysis, find the recurrence relations satisfied by $H_i(s)$ for $i \geqslant 2$ and $i \leqslant -2$, and for $i = -1$, $i = 0$, $i = 1$.

(b) Find $H_i(s)$ for $i \geqslant 1$, $i = 0$, and $i \leqslant -1$.

Hint. Look for a solution of the form

$$H_i(s) = C(s)\alpha^i(s) \text{ for } i \geqslant 1 \text{ and } i \leqslant -1.$$

(c) Consider the probability generating function $G_i(s)$ of the first return time to ⓪, defined as

$$G_i(s) := \mathbb{E}\big[s^{T_0^1} \,\big|\, S_0 = i\big], \qquad i \in \mathbb{Z}, \quad -1 \leqslant s \leqslant 1.$$

Using conditioning based on T_0^1, find a relation between $G_i(s)$, $H_i(s)$ and $H_0(s)$ for $i \geq 2$ and $i \leq -2$, and for $i = -1$, $i = 0$, $i = 1$.

(d) Find $G_i(s)$ for $i \geq 1$, $i = 0$, and $i \leq -1$.

(e) Find the probability $\mathbb{P}(T_0^1 < \infty \mid S_0 = i)$ of hitting state $\textcircled{0}$ in finite time after starting from state $\textcircled{i}$.

(f) Find the mean number of visits $\mathbb{E}[R_0 \mid S_0 = i]$ to state $\textcircled{0}$ after starting from state $\textcircled{i}$, $i \in \mathbb{Z}$.

Problem 4.5 Time spent above zero by a random walk. Consider the *symmetric* random walk $(S_n)_{n \geq 0}$ started at $S_0 = 0$ on $\mathbb{S} = \mathbb{Z}$. We let

$$T_{2n}^+ := 2 \sum_{r=1}^{n} \mathbb{1}_{\{S_{2r-1} \geq 1\}}$$

denote an even estimate of the time spent strictly above the level 0 by the random walk between time 0 and time $2n$. We also let

$$T_0 := \inf\{n \geq 1 : S_n = 0\}$$

denote an even estimate of the time of first return of $(S_n)_{n \geq 0}$ to $\textcircled{0}$.

(a) Compute $\mathbb{P}(S_{2n} = 2k)$ for $k = 0, 1, \ldots, n$.

(b) Show the *convolution equation*

$$\mathbb{P}(S_{2n} = 0) = \sum_{r=1}^{n} \mathbb{P}(T_0 = 2r)\mathbb{P}(S_{2n-2r} = 0), \qquad n \geq 1.$$

(c) By partitioning the event $\{T_{2n}^+ = 2k\}$ according to all possible times $2r = 2, 4, \ldots, 2n$ of *first* return to state $\textcircled{0}$ until time $2n$, show the convolution equation

$$\mathbb{P}(T_{2n}^+ = 2k) = \sum_{r=1}^{n} \mathbb{P}(T_0 = 2r, T_{2n}^+ = 2k)$$

$$= \frac{1}{2} \sum_{r=1}^{k} \mathbb{P}(T_0 = 2r)\mathbb{P}(T_{2n-2r}^+ = 2k - 2r)$$

$$+ \frac{1}{2} \sum_{r=1}^{n-k} \mathbb{P}(T_0 = 2r)\mathbb{P}(T_{2n-2r}^+ = 2k), \qquad n \geq 1.$$

(d) Show that

$$\mathbb{P}(T_{2n}^+ = 2k) = \mathbb{P}(S_{2k} = 0)\mathbb{P}(S_{2n-2k} = 0), \qquad 0 \leqslant k \leqslant n,$$

solves the convolution equation of Question (c).

(e) Using the Stirling approximation $n! \simeq (n/e)^n \sqrt{2\pi n}$ as n tends to infinity, compute the limit

$$\lim_{n \to \infty} \mathbb{P}\big(T_{2n}^+/(2n) \leqslant x\big) = \lim_{n \to \infty} \sum_{0 \leqslant k \leqslant nx} \mathbb{P}\big(T_{2n}^+/(2n) = k/n\big),$$

and find the limiting distribution of $T_{2n}^+/(2n)$ as n tends to infinity.

Problem 4.6 Consider a sequence $(X_n)_{n \geqslant 1}$ of independent random variables on $\{1, \ldots, d\}$ with same distribution $\pi = (\pi_1, \ldots, \pi_d)$. In what follows,

$$f : \{1, \ldots, d\} \to \mathbb{R}$$

denotes any function such that $\|f\|_\infty \leqslant 1$ and $\mathbb{E}[f(X_n)] = 0$, $n \geqslant 1$, and we let

$$\lambda_0(\alpha) := \sum_{l=1}^{d} \pi_l e^{\alpha f(l)}, \qquad \alpha \geqslant 0.$$

(a) Show that for any $\alpha \in \mathbb{R}$ we have

$$\mathbb{E}\left[\exp\left(\alpha \sum_{l=1}^{n} f(X_l)\right)\right] = (\lambda_0(\alpha))^n, \qquad n \geqslant 0.$$

(b) Show that for any $\alpha \in \mathbb{R}$ and $\gamma > 0$ we have

$$\mathbb{P}\left(\frac{1}{n} \sum_{l=1}^{n} f(X_l) \geqslant \gamma\right) \leqslant e^{-n(\alpha\gamma - \log \lambda_0(\alpha))}, \qquad n \geqslant 1.$$

Hint. Use the Chernoff argument.

(c) Show that

$$\lambda_0(\alpha) = 1 + \sum_{l=1}^{d} \pi_l (e^{\alpha f(l)} - \alpha f(l) - 1), \qquad \alpha \geqslant 0.$$

(d) Show that

$$\lambda_0(\alpha) \leqslant 1 + \frac{\alpha^2}{1 - \alpha}, \qquad \alpha \in [0, 1).$$

(e) Show that for any $\alpha \in [0, 1)$ and $\gamma > 0$ we have

$$\mathbb{P}\left(\frac{1}{n}\sum_{l=1}^{n} f(X_l) \geqslant \gamma\right) \leqslant e^{-n(\alpha\gamma - \frac{\alpha^2}{1-\alpha})}, \qquad n \geqslant 1.$$

(f) Find the value of $\alpha \in [0, 1)$ which maximizes $\alpha\gamma - \alpha^2/(1 - \alpha)$.
(g) Show that for all $\gamma > 0$ and $n \geqslant 1$ we have

$$\mathbb{P}\left(\frac{1}{n}\sum_{l=1}^{n} f(X_l) \geqslant \gamma\right) \leqslant e^{-n\gamma^2/6}.$$

Problem 4.7 Consider a sequence $(X_n)_{n\geqslant 0}$ of independent identically distributed random variables with distribution $\pi = (\pi_1, \ldots, \pi_d)$ on $\{1, \ldots, d\}$. Our goal is to estimate the distribution π using the estimator $\widehat{\pi}_j(n) := \frac{1}{n}\sum_{k=1}^{n} \mathbf{1}_{\{X_k=j\}}$, $j = 1, \ldots, d$.

(a) Show that $\mathbb{E}\left[\sum_{j=1}^{d}\left|\frac{1}{n}\sum_{k=1}^{n}\mathbf{1}_{\{X_k=j\}} - \pi_j\right|\right] \leqslant \sqrt{\frac{d}{n}}, i = 1, \ldots, d.$

(b) Show that for any $n \geqslant 1$, the function $f_n : \mathbb{R}^n \to \mathbb{R}$ defined by

$$(x_1, \ldots, x_n) \mapsto \sum_{j=1}^{d}\left|\frac{1}{n}\sum_{k=1}^{n}\mathbf{1}_{\{x_k=j\}} - \pi_j\right|$$

satisfies the bounded differences property with $c_i = 2/n$, $i = 1, \ldots, n$, i.e.

$$\operatorname*{Sup}_{y\in\mathbb{R}} |f(x_1, \ldots, x_n) - f(x_1, \ldots, x_{i-1}, y, x_{i+1}, \ldots, x_n)| \leqslant c_i, \quad x_1, \ldots, x_n \in \mathbb{R}.$$

(c) Based on the results of Questions (a)–(b) and McDiarmid's inequality

$$\mathbb{P}(f(X_1, \ldots, X_n) - \mathbb{E}[f(X_1, \ldots, X_n)] \geqslant \varepsilon) \leqslant \exp\left(-\frac{2\varepsilon^2}{\sum_{i=1}^{n} c_i^2}\right),$$

show that for all $i = 1, \ldots, d$ we have

$$\mathbb{P}\left(\sum_{j=1}^{d}\left|\frac{1}{n}\sum_{k=1}^{n}\mathbf{1}_{\{X_k=j\}} - \pi_j\right| > \varepsilon\right) \leqslant \exp\left(-\frac{n}{2}\,\mathrm{Max}\left(0, \varepsilon - \sqrt{\frac{d}{n}}\right)^2\right).$$

(d) Show that if $n \geqslant 4d/\varepsilon^2$, then we have $\mathbb{P}\left(\sum_{j=1}^{d}\left|\widehat{\pi}_j(n) - \pi_j\right| > \varepsilon\right) \leqslant e^{-n\varepsilon^2/8}$.

(e) Show that there is a constant $c > 0$ such that for any $\varepsilon, \delta \in (0, 1)$ we have

$$\mathbb{P}\left(\mathrm{Max}_{j=1,\ldots,d}\left|\widehat{\pi}_j(n) - \pi_j\right| \leqslant \varepsilon\right) \geqslant 1 - \delta,$$

for any $n > c\log(1/\delta)/\varepsilon^2$.

Chapter 5
Cookie-Excited Random Walks

In this chapter we consider random walks in a cookie environment, also called excited random walks (ERWs), which are not Markovian and are used in physics and biology, to model the behavior of e.g. primitive organisms. Random walks in a random environment can be used for the understanding of macroscopic phenomena by rescaling, based on the modeling of random trajectories at a microscopic level.

5.1 Hitting Times and Probabilities

We assume that the state space $S := \{0, 1, 2, \ldots\}$ is equipped with "cookies" at the locations (n), $n \geqslant 1$, and consider a random walk $(S_n)_{n \geqslant 0}$ which moves with probabilities (p, q) of going up and down in the absence of cookies. The random walk starts from state (0), which has no cookie. After hitting state (0) it can *rebound* to state (1) with probability p, or return to state (0) with probability q, see Fig. 5.1.

When the random walk encounters a cookie, its behavior becomes modified and the next state is chosen with the probabilities $\widetilde{p} \in [0, 1]$ and $\widetilde{q} := 1 - \widetilde{p}$ of moving up, resp. down, independently of the past. Every encountered cookie is eaten by the organism, and when the random walk reaches an empty spot it continues with the probabilities (p, q) of moving up or down. The random walk is *attracted* by the cookies when $\widetilde{p} > 1/2$, and *repulsed* when $\widetilde{p} < 1/2$.

The cookie random walk does not have the Markov property when $\widetilde{p} \neq p$ because in this case the transition probabilities at a given state may depend on the past behavior of the chain starting from time 1. On the other hand, the cookie random walk has the Markov property when $\widetilde{p} = p$ because in this case it coincides with the usual symmetric random walk with independent increments.

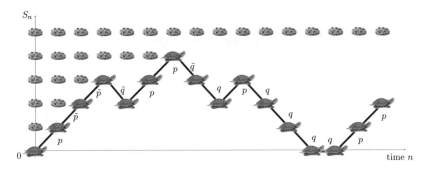

Fig. 5.1 Random walk with cookies

Hitting Probabilities

For any $x \in \mathbb{N}$, let T_x^r denote the first return time

$$T_x^r := \inf\{n \geqslant 1 \ : \ S_n = x\}.$$

Proposition 5.1 *The hitting probability* $\mathbb{P}\big(T_x^r < T_0^r \mid S_0 = 0\big)$ *takes the form*

$$\mathbb{P}\big(T_x^r < T_0^r \mid S_0 = 0\big) = p \prod_{l=1}^{x-1} (1 - f(l)), \quad x \geqslant 1,$$

with

$$f(l) = \frac{(q-p)\widetilde{q}}{(1 - (p/q)^{l+1})q^2} \leqslant \widetilde{q} \leqslant 1, \qquad l \geqslant 1,$$

when $p \neq q$, and

$$f(l) = \frac{2\widetilde{q}}{l+1} \leqslant \widetilde{q} \leqslant 1, \qquad l \geqslant 1,$$

when $p = q = 1/2$. Note that when $\widetilde{q} = 1$ we have $\mathbb{P}\big(T_x^r < T_0^r \mid S_0 = 0\big) = 0$, $x \geqslant 2$, as $f(2) = 0$.

Proof In this proof, $\mathbb{P}(\cdot \mid S_0 = \widehat{x})$ denotes the conditional probability given that a cookie has just been eaten at state $\widehat{x}$. Assume that the random walk has just eaten a cookie at state $x \geqslant 1$, after eating all cookies at states $1, 2, \ldots, x - 1$. If $p \neq q$, by first step analysis, the probability of reaching $\boxed{x + 1}$ before reaching $\textcircled{0}$ is given from (1.43) as

$$
\begin{aligned}
\mathbb{P}\big(T^r_{x+1} < T^r_0 \mid S_0 = \widehat{x}\big) &= \widetilde{p}\,\mathbb{P}\big(T^r_{x+1} < T^r_0 \mid S_1 = \widehat{x+1}\big) \\
&\quad + \widetilde{q}\,\mathbb{P}\big(T^r_{x+1} < T^r_0 \mid S_1 = x - 1\big) \\
&= \widetilde{p} + \widetilde{q}\,\frac{1 - (q/p)^{x-1}}{1 - (q/p)^{x+1}} \\
&= 1 - \frac{(p - q)(q/p)^x \widetilde{q}}{(1 - (q/p)^{x+1})pq} \\
&= 1 + \frac{(p - q)\widetilde{q}}{(1 - (p/q)^{x+1})q^2}
\end{aligned} \tag{5.1}
$$

If $p = q = 1/2$, from (1.44) we have

$$
\mathbb{P}\big(T^r_{x+1} < T^r_0 \mid S_0 = \widehat{x}\big) = \widetilde{p} + \widetilde{q}\,\frac{x - 1}{x + 1} = 1 - \frac{2\widetilde{q}}{x + 1}, \quad x \geqslant 1, \tag{5.2}
$$

since the probability for a symmetric random walk to reach state $\boxed{x + 1}$ before hitting state $\textcircled{0}$ starting from $\textcircled{k}$ is $k/(x + 1)$, see formula (1.44) p. 38. We have $\mathbb{P}\big(T^r_1 < T^r_0 \mid S_0 = 0\big) = p$ and by the (strong) Markov property, by reasoning inductively on the transitions from state $\textcircled{0}$ to state $\textcircled{1}$, then from state $\textcircled{2}$ to state $\textcircled{2}$, etc, up to state $\textcircled{x}$, we find

$$
\begin{aligned}
\mathbb{P}\big(T^r_x < T^r_0 \mid S_0 = 0\big) &= \mathbb{P}\big(T^r_1 < T^r_0 \mid S_0 = 0\big) \prod_{l=1}^{x-1} \mathbb{P}\big(T^r_{l+1} < T^r_0 \mid S_0 = \widehat{l}\big) \\
&= p \prod_{l=1}^{x-1} \left(1 - \frac{(p - q)(q/p)^l \widetilde{q}}{(1 - (q/p)^{l+1})pq}\right) \\
&= p \prod_{l=2}^{x} \left(1 - \frac{(p - q)(q/p)^l \widetilde{q}}{(1 - (q/p)^l)q^2}\right)
\end{aligned}
$$

if $p \neq q$, and

$$\mathbb{P}(T_x^r < T_0^r \mid S_0 = 0) = \mathbb{P}(T_1^r < T_0^r \mid S_0 = 0) \prod_{l=1}^{x-1} \mathbb{P}(T_{l+1}^r < T_0^r \mid S_0 = \hat{l})$$

$$= \frac{1}{2} \prod_{l=1}^{x-1} \left(1 - \frac{2\tilde{q}}{l+1}\right)$$

$$= \frac{1}{2} \prod_{l=2}^{x} \left(1 - \frac{2\tilde{q}}{l}\right), \qquad x \geqslant 1,$$

if $p = q = 1/2$. $\square$

The result of Proposition 5.1 can also be written as

$$\mathbb{P}(T_x^r < T_0^r \mid S_0 = 0) = p \prod_{l=2}^{x} \left(1 - \frac{(q-p)\tilde{q}}{(1-(p/q)^l)q^2}\right) \tag{5.3}$$

if $p \neq q$, and

$$\mathbb{P}(T_x^r < T_0^r \mid S_0 = 0) = \frac{1}{2} \prod_{l=2}^{x} \left(1 - \frac{2\tilde{q}}{l}\right), \qquad x \geqslant 1, \tag{5.4}$$

if $p = q = 1/2$.

For $x = 2$, (5.3) and (5.4) show that

$$\mathbb{P}(T_2^r < T_0^r \mid S_0 = 0) = p\left(1 - \frac{(q-p)\tilde{q}}{(1-(p/q)^2)q^2}\right)$$

$$= p\frac{(1-(p/q)^2)q^2 - (q-p)\tilde{q}}{(1-(p/q)^2)q^2}$$

$$= p\frac{q^2 - p^2 - (q-p)\tilde{q}}{p^2 - q^2}$$

$$= p\tilde{p}$$

when $p \neq q$, and

$$\mathbb{P}(T_2^r < T_0^r \mid S_0 = 0) = \frac{\tilde{p}}{2}$$

when $p = q = 1/2$. In particular, when $\tilde{q} = 1$ we have $\mathbb{P}(T_x^r < T_0^r \mid S_0 = 0) = 0$ for all $x \geqslant 2$.

For all $x \geqslant 1$, we also have[1]

$$\mathbb{P}(T_x^r < T_0^r \mid S_0 = 0) = p \exp\left(\sum_{l=2}^{x} \log\left(1 + \frac{(p-q)\widetilde{q}}{(1-(p/q)^l)q^2}\right)\right) \tag{5.5}$$

if $p \neq q$, and

$$\mathbb{P}(T_x^r < T_0^r \mid S_0 = 0) = \frac{1}{2} \exp\left(\sum_{l=2}^{x} \log\left(1 - \frac{2\widetilde{q}}{l}\right)\right), \quad x \geqslant 1, \tag{5.6}$$

if $p = q = 1/2$, where "log" denotes the *natural logarithm* "ln".

5.2 Recurrence

The symmetric case $p = q = 1/2$ is treated in § 3.3 of Antal and Redner (2005), see also § 2 of Benjamini and Wilson (2003). Excited random walks on $\mathbb{Z}^d$ are treated in Benjamini and Wilson (2003), where it is shown that excited symmetric random walks are transient if and only if $d \geqslant 2$. The next result shows in particular that the reflected random walk of Sect. 4.3 is recurrent when $p = \widetilde{p} \leqslant 1/2$, and transient when $p = \widetilde{p} > 1/2$.

Proposition 5.2

(a) When $p \leqslant 1/2$, the cookie-excited random walk is recurrent for all $\widetilde{p} \in [0, 1)$.
(b) When $p > 1/2$, the cookie-excited random walk is transient for all $\widetilde{p} \in (0, 1]$.

Proof We note that the sequence $(T_x)_{x \geqslant 1}$ is strictly increasing and $\lim_{x \to \infty} T_x = +\infty$ almost surely since $x \leqslant T_x^r < T_{x+1}^r$, $x \geqslant 1$. Hence, we have

$$\{T_0^r < \infty\} = \bigcup_{x \geqslant 1} \{T_0^r < T_x^r\},$$

and therefore

$$\mathbb{P}(T_0^r < \infty \mid S_0 = 0) = \mathbb{P}\left(\bigcup_{x \geqslant 1} \{T_0^r < T_x^r\} \;\middle|\; S_0 = 0\right)$$

$$= \lim_{x \to \infty} \mathbb{P}(T_0^r < T_x^r \mid S_0 = 0)$$

---

[1] We use the convention $\sum_{k=2}^{1} a_k = 0$ for any sequence (a_k).

$$= \lim_{x \to \infty} \left(1 - \mathbb{P}(T_x^r \leqslant T_0^r \mid S_0 = 0) \right)$$

$$= 1 - \lim_{x \to \infty} \mathbb{P}(T_x^r < T_0^r \mid S_0 = 0) \qquad (5.7)$$

since $\mathbb{P}(T_x^r = T_0^r \mid S_0 = 0) = 0$.

(a) Case $p \in [0, 1/2]$.

    (i) Case $p = q = 1/2$. Using again the inequality $\log(1 + z) \leqslant z$ for $z > -1$, (Fig. 5.2), by (5.6) we have

$$\int_2^x \log \left(1 - \frac{2\widetilde{q}}{y} \right) dy \leqslant \sum_{l=2}^x \log \left(1 - \frac{2\widetilde{q}}{l} \right)$$

$$\leqslant \int_2^{x+1} \log \left(1 - \frac{2\widetilde{q}}{y} \right) dy$$

$$\leqslant -2\widetilde{q} \int_2^{x+1} \frac{1}{y} dy$$

$$= -2\widetilde{q} \log \frac{x+1}{2},$$

hence

$$\mathbb{P}(T_x^r < T_0^r \mid S_0 = 0) = \frac{1}{2} \exp \left(\sum_{l=2}^x \log \left(1 - \frac{2\widetilde{q}}{l} \right) \right)$$

$$\leqslant \exp \left(-2\widetilde{q} \log \frac{x}{2} \right)$$

$$= \left(\frac{x}{2} \right)^{-2\widetilde{q}}, \qquad x \geqslant 2,$$

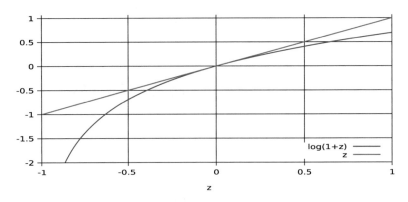

Fig. 5.2 Log function

hence

$$\lim_{x \to \infty} \mathbb{P}(T_x^r < T_0^r \mid S_0 = 0) = \lim_{x \to \infty} \left(\frac{x}{2}\right)^{-2\widetilde{q}} = 0,$$

when $\widetilde{q} \in (0, 1]$, and we conclude from (5.7).

(ii) Case $p \in [0, 1/2)$. By Proposition 4.13 and the fact that $\mathbb{P}(B \cap A) = \mathbb{P}(B)$ when $\mathbb{P}(A) = 1$, we note that $\mathbb{P}(T_x^r < \infty \mid S_0 = 0) = 1$ for all $x \geqslant 1$. Next, for any $p < 1/2 < q$ and $\widetilde{q} \in (0, 1]$ and any $\varepsilon > 0$, there exists $l_0 \geqslant 1$ large enough such that

$$\frac{(q - p)\widetilde{q}}{q^2} - \varepsilon < \frac{(p - q)(q/p)^l \widetilde{q}}{(1 - (q/p)^l) q^2} < \frac{(q - p)\widetilde{q}}{q^2} + \varepsilon, \qquad l \geqslant l_0,$$

and by a comparison argument between integrals and series, we find

$$(x - l_0) \log\left(1 - \frac{(q - p)\widetilde{q}}{q^2} - \varepsilon\right)$$

$$\leqslant \int_{l_0}^{x} \log\left(1 - \frac{(p - q)(q/p)^y \widetilde{q}}{(1 - (q/p)^y) q^2}\right) dy$$

$$\leqslant \sum_{l=l_0}^{x} \log\left(1 - \frac{(p - q)(q/p)^l \widetilde{q}}{(1 - (q/p)^l) q^2}\right)$$

$$\leqslant \int_{l_0}^{x+1} \log\left(1 - \frac{(p - q)(q/p)^y \widetilde{q}}{(1 - (q/p)^y) q^2}\right) dy$$

$$\leqslant (x + 1 - l_0) \log\left(1 - \frac{(q - p)\widetilde{q}}{q^2} + \varepsilon\right),$$

hence by (5.5) we obtain

$$C_{l_0} \left(1 - \frac{(q - p)\widetilde{q}}{q^2} + \varepsilon\right)^{x - l_0}$$

$$\leqslant \mathbb{P}(T_x^r < T_0^r \mid S_0 = 0)$$

$$\leqslant C_{l_0} \left(1 - \frac{(q - p)\widetilde{q}}{q^2} + \varepsilon\right)^{x + 1 - l_0}, \qquad x \geqslant l_0,$$

for some $C_{l_0} > 0$, showing that $\lim_{x\to\infty} \mathbb{P}\big(T_x^r < T_0^r \mid S_0 = 0\big) = 0$ provided that $\widetilde{q} > 0$.[2]

(b) Case $p \in (1/2, 1]$. By Proposition 5.1, we have

$$
\lim_{x\to\infty} \mathbb{P}(T_x^r < T_0^r \mid S_0 = 0) = p \lim_{x\to\infty} \prod_{l=2}^{x} \left(1 - \frac{(p-q)(q/p)^l\widetilde{q}}{(1-(q/p)^l)q^2}\right)
$$

$$
= p \prod_{l=2}^{\infty} \left(1 - \frac{(p-q)(q/p)^l\widetilde{q}}{(1-(q/p)^l)q^2}\right)
$$

$$
= p \exp\left(\sum_{l=2}^{\infty} \log\left(1 - \frac{(p-q)(q/p)^l\widetilde{q}}{(1-(q/p)^l)q^2}\right)\right).
$$

When $\widetilde{q} = 1$, since $(p-q)(q/p)^2/((1-(q/p)^2)q^2) = 1$ we have $\mathbb{P}(T_x^r < T_0^r \mid S_0 = 0) = 0$, $x \geqslant 2$, hence $\lim_{x\to\infty} \mathbb{P}(T_x^r < T_0^r \mid S_0 = 0) = 0$, and the random walk is recurrent in this case. On the other hand, when $\widetilde{q} \in [0, 1)$ we have

$$
\widetilde{q}\,\frac{(p-q)(q/p)^l}{(1-(q/p)^l)q^2} \leqslant \widetilde{q} < 1, \qquad l \geqslant 2,
$$

and

$$
\log\left(1 - \frac{(p-q)(q/p)^l\widetilde{q}}{(1-(q/p)^l)q^2}\right) \simeq -\frac{(p-q)(q/p)^l\widetilde{q}}{(1-(q/p)^l)q^2} \simeq -\frac{(p-q)\widetilde{q}}{q^2}\left(\frac{q}{p}\right)^l.
$$

Hence, as l tends to infinity, we obtain

$$
-\infty < \sum_{l=2}^{\infty} \log\left(1 - \frac{(p-q)(q/p)^l\widetilde{q}}{(1-(q/p)^l)q^2}\right) \leqslant 0
$$

by the limit comparison test, which yields $\lim_{x\to\infty} \mathbb{P}(T_x^r < T_0^r \mid S_0 = 0) > 0$, hence by (5.7) we find

$$
\mathbb{P}(T_0^r < \infty \mid S_0 = 0) = 1 - \lim_{x\to\infty} \mathbb{P}(T_x^r < T_0^r \mid S_0 = 0) < 1.
$$

$\square$

The next proposition provides more precise estimates of $\mathbb{P}(T_0 = \infty \mid S_0 = 0)$ in the transient case $p > 1/2$.

---

[2] Note that $(q - p)/q^2 < 1$ when $q \in [1/2, 1)$.

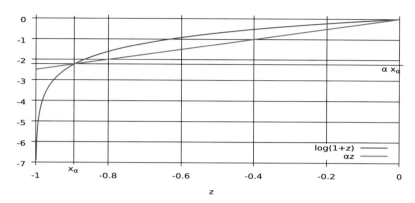

Fig. 5.3 log function

Proposition 5.3 *When $p > 1/2$ we have*

$$p\widetilde{p}\left(1 - \widetilde{q}\frac{q}{p}\right)^{p/(p-q)} \leqslant \mathbb{P}(T_0^r = \infty \mid S_0 = 0) \leqslant p\widetilde{p}.$$

In particular, $\mathbb{P}(T_0^r = \infty \mid S_0 = 0)$ is strictly positive if and only if $\widetilde{p} \in (0, 1]$, i.e. $\widetilde{q} \in [0, 1)$.

Proof Let $\alpha > 1$, and consider the inequality

$$\alpha z \leqslant \log(1 + z) \leqslant 0, \qquad x_\alpha \leqslant z \leqslant 0,$$

with $\alpha x_\alpha = \log(1 + x_\alpha)$, see Fig. 5.3.
We have

$$-\alpha\widetilde{q}\frac{q^l}{p^l} \leqslant \log\left(1 - \widetilde{q}\frac{q^l}{p^l}\right),$$

provided that

$$x_\alpha \leqslant -\widetilde{q}\frac{q^l}{p^l} \leqslant 0,$$

i.e.

$$l \geqslant \frac{\log(-x_\alpha/\widetilde{q})}{\log(q/p)} = 1.$$

Hence, choosing $x_\alpha := -q\widetilde{q}/p$, we have

$$\alpha = -\frac{p}{q\widetilde{q}} \log\left(1 - \widetilde{q}\frac{q}{p}\right) \geqslant 1.$$

Hence, using the relation

$$\mathbb{P}(T_x^r < T_0^r \mid S_0 = 0) = p \exp\left(\sum_{l=2}^{x} \log\left(1 - \frac{(p-q)(q/p)^l \widetilde{q}}{(1-(q/p)^l)q^2}\right)\right), \quad x \geqslant 1,$$

from Proposition 5.1 and the relation $(p-q)(q/p)^l/((1-(q/p)^l)q^2) = 1$ for $l = 2$, we have

$$\log \mathbb{P}(T_x^r < T_0^r \mid S_0 = 0) = \log p + \sum_{l=2}^{x} \log\left(1 - \frac{(p-q)(q/p)^l \widetilde{q}}{(1-(q/p)^l)q^2}\right)$$

$$\geqslant \log(1-\widetilde{q}) + \log p + \sum_{l=3}^{x} \log\left(1 - \frac{(p-q)(q/p)^l \widetilde{q}}{(1-(q/p)^2)q^2}\right)$$

$$= \log(1-\widetilde{q}) + \log p + \sum_{l=3}^{x} \log\left(1 - \widetilde{q}\left(\frac{q}{p}\right)^{l-2}\right)$$

$$\geqslant \log((1-\widetilde{q})p) - \alpha\widetilde{q} \sum_{l=1}^{x} \left(\frac{q}{p}\right)^l$$

$$= \log((1-\widetilde{q})p) - \alpha\widetilde{q}\frac{q}{p} \sum_{k=0}^{\infty} \left(\frac{q}{p}\right)^k$$

$$= \log((1-\widetilde{q})p) - \alpha\frac{\widetilde{q}q}{p-q}, \quad x \geqslant 2,$$

hence

$$\lim_{x\to\infty} \mathbb{P}(T_x^r < T_0^r \mid S_0 = 0) \geqslant p(1-\widetilde{q}) \exp\left(-\frac{\alpha\widetilde{q}q}{p-q}\right),$$

and

$$1 - p\widetilde{p} = q + p\widetilde{q} \leqslant \mathbb{P}(T_0^r < \infty \mid S_0 = 0) \leqslant 1 - p\widetilde{p}\left(1 - \widetilde{q}\frac{q}{p}\right)^{p/(p-q)} < 1.$$

We note that the bound becomes an equality at $\widetilde{p} = 0$ and $\widetilde{p} = 1$.   □

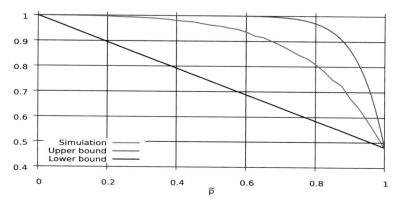

Fig. 5.4 Upper and lower bounds on $\mathbb{P}(T_0^r < \infty \mid S_0 = 0)$ with $p = 0.52$ on $[0, 1]$

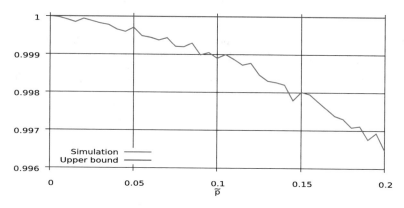

Fig. 5.5 Upper and lower bounds on $\mathbb{P}(T_0^r < \infty \mid S_0 = 0)$ with $p = 0.52$ on $[0, 0.2]$

The C code_11_page_133.c is used to plot Figs. 5.4 and 5.5 with 10,000 samples and tmax= $100,000$, see also the cookie_recurrence_page_133.ipynb IPython notebook, which provides a Monte Carlo estimate of the probability of return to zero within a given time.

5.3 Mean Hitting Times

Recall that the mean time needed by the random walk to reach state ① after starting from state ⓪ can be computed in at least three different ways.

(i) By first step analysis. We have

$$\mathbb{E}\big[T_1^r \,\big|\, S_0 = 0\big] = p \times 1 + q\big(1 + \mathbb{E}\big[T_1^r \,\big|\, S_0 = 0\big]\big),$$

hence

$$\mathbb{E}\big[T_1^r \,|\, S_0 = 0\big] = \frac{1}{p}. \tag{5.8}$$

(ii) By pathwise analysis. We have

$$\mathbb{E}\big[T_1^r \,|\, S_0 = 0\big] = p \sum_{k \geqslant 1} k q^{k-1} = \frac{p}{(1-q)^2} = \frac{1}{p}.$$

(iii) By applying Proposition 4.14 with $L = 1$ and $k = 0$, which recovers

$$\mathbb{E}\big[T_1^r \,|\, S_0 = 0\big] = \frac{1}{p-q} + \frac{q}{(p-q)^2}\left(\frac{q}{p} - 1\right) = \frac{1}{p}.$$

When $p = q = 1/2$ we find $\mathbb{E}\big[T_1^r \,|\, S_0 = 0\big] = 2$. We can check that the result of the next proposition is consistent with that of by Proposition 4.14 when $\widetilde{q} = q$.

Proposition 5.4 Let $x \geqslant 1$. The mean time to reach state $\textcircled{x}$ starting from $\textcircled{0}$ is given by

$$\mathbb{E}\big[T_x^r \,|\, S_0 = 0\big] = \frac{q - \widetilde{q}}{p} + \left(1 + \frac{2\widetilde{q}}{p-q}\right)x + \frac{\widetilde{q}}{(p-q)^2}\left(\left(\frac{q}{p}\right)^x - 1\right),$$

when $p \neq q$, and by

$$\mathbb{E}\big[T_x^r \,|\, S_0 = 0\big] = 1 - 2\widetilde{q} + x + 2\widetilde{q}x^2, \quad x \geqslant 1,$$

when $p = q = 1/2$.

Proof Assume that a cookie has just been eaten at state $x \geqslant 1$, after eating all cookies at states $1, 2, \ldots, x - 1$.

(i) When $p \neq q$, by Proposition 4.14 applied to $k = x - 1$ and $L = x + 1$ and first step analysis, we find that the mean time to reach the next cookie at state $\boxed{x + 1}$ is given by

$$\mathbb{E}\big[T_{x+1}^r \,|\, S_0 = \widehat{x}\big]$$

$$= \widetilde{p} + \widetilde{q}\left(1 + \frac{x+1-(x-1)}{p-q} + \frac{q}{(p-q)^2}\left(\left(\frac{q}{p}\right)^{x+1} - \left(\frac{q}{p}\right)^{x-1}\right)\right)$$

$$= 1 + \widetilde{q}\left(\frac{2}{p-q} + \frac{q}{(p-q)^2}\left(\left(\frac{q}{p}\right)^{x+1} - \left(\frac{q}{p}\right)^{x-1}\right)\right)$$

$$= 1 + \frac{2\widetilde{q}}{p - q} - \frac{\widetilde{q}}{(p - q)p} \left(\frac{q}{p}\right)^x$$

$$= 1 + \frac{\widetilde{q}}{p - q} \left(2 - \frac{1}{p} \left(\frac{q}{p}\right)^x\right).$$

Next, we proceed by summing (5.8) and the above expression, as follows:

$$\mathbb{E}\left[T_x^r \mid S_0 = 0\right] = \sum_{k=0}^{x-1} \mathbb{E}\left[T_{k+1}^r \mid S_0 = \widehat{k}\right]$$

$$= \mathbb{E}\left[T_1^r \mid S_0 = 0\right] + \sum_{k=1}^{x-1} \left(1 + \frac{2\widetilde{q}}{p - q} - \frac{\widetilde{q}}{(p - q)p} \left(\frac{q}{p}\right)^k\right)$$

$$= \frac{1}{p} + \left(1 + \frac{2\widetilde{q}}{p - q}\right)(x - 1) - \frac{\widetilde{q}q}{p^2(p - q)} \sum_{k=0}^{x-2} \left(\frac{q}{p}\right)^k$$

$$= \frac{1}{p} + \left(1 + \frac{2\widetilde{q}}{p - q}\right)(x - 1) - \frac{\widetilde{q}q}{p^2(p - q)} \frac{1 - (q/p)^{x-1}}{1 - q/p}$$

$$= \frac{1}{p} + \left(1 + \frac{2\widetilde{q}}{p - q}\right)(x - 1) - \frac{\widetilde{q}q}{(p - q)^2 p}(1 - (q/p)^{x-1}), \quad x \geqslant 1.$$

(ii) Similarly, when $p = q = 1/2$, by Proposition 4.14 applied to $k = x - 1$ and $L = x + 1$ and first step analysis, we find

$$\mathbb{E}\left[T_{x+1}^r \mid S_0 = \widehat{x}\right] = \widetilde{p} + (1 + (x + 1 + (x - 1) + 1)(x + 1 - (x - 1)))\widetilde{q}$$

$$= \widetilde{p} + (1 + 4x + 2)\widetilde{q}$$

$$= 1 + 2(2x + 1)\widetilde{q}.$$

Next, we proceed by summing (5.8) and the above result, as follows:

$$\mathbb{E}\left[T_x^r \mid S_0 = 0\right] = \sum_{k=0}^{x-1} \mathbb{E}\left[T_{k+1}^r \mid S_0 = \widehat{k}\right]$$

$$= 2 + \sum_{k=1}^{x-1} (1 + (4k + 2)\widetilde{q})$$

$$= 2 + (1 + 2\widetilde{q})(x - 1) + 4\widetilde{q} \sum_{k=1}^{x-1} k$$

$$= 2 + (1 + 2\widetilde{q})(x - 1) + 2\widetilde{q}x(x - 1)$$

$$= 2 + (1 + 2\widetilde{q})x - (1 + 2\widetilde{q}) + 2\widetilde{q}x^2 - 2\widetilde{q}x$$
$$= 1 - 2\widetilde{q} + x + 2\widetilde{q}x^2, \qquad x \geqslant 1.$$

□

Letting $p := (1 + \varepsilon)/2$ and $q := (1 - \varepsilon)/2$ we check that the following equivalences hold as ε tends to zero:

$$\frac{q - \widetilde{q}}{p} + \left(1 + \frac{2\widetilde{q}}{p - q}\right)x + \frac{\widetilde{q}}{(p - q)^2}\left(\left(\frac{q}{p}\right)^x - 1\right)$$

$$\simeq 1 - 2\widetilde{q} + x + \frac{2\widetilde{q}}{\varepsilon}x + \frac{\widetilde{q}}{\varepsilon}\left((1 - \varepsilon)^x(1 + \varepsilon)^{-x} - 1\right)$$

$$\simeq 1 - 2\widetilde{q} + x + \frac{2\widetilde{q}}{\varepsilon}x + \frac{\widetilde{q}}{\varepsilon}\left(-2\varepsilon x + \varepsilon^2 x(x - 1) + \varepsilon^2 x(x + 1)\right)$$

$$\simeq 1 - 2\widetilde{q} + x + \frac{2\widetilde{q}}{\varepsilon}x + \frac{\widetilde{q}}{\varepsilon^2}(-2\varepsilon x + 2\varepsilon^2 x^2)$$

$$\simeq 1 - 2\widetilde{q} + x + 2\widetilde{q}x^2, \qquad x \geqslant 1.$$

Remark 5.5 One can also show that when $p = q = 1/2$, for all $\widetilde{q} < 1$ the mean return time $\mathbb{E}\big[T_0^r \mid S_0 = 0\big]$ to state $\textcircled{0}$ is *infinite*, showing that the cookie random walk is *null recurrent*, see page 2563 of Antal and Redner (2005).

5.4 Count of Cookies Eaten

Recall that the random walk $(S_n)_{n \geqslant 0}$ with cookies on $\{1, 2, 3, \ldots\}$ is symmetric in the absence of cookies, and restarts with probabilities p and $q = 1 - p$ of moving up, resp. down, when it encounters a cookie, where $p \in [0, 1)$. The random walk starts at state $\textcircled{0}$, which is empty of cookie.

For any $x \geqslant 1$, let T_x^r denote the first return time

$$T_x^r := \inf\{n \geqslant 1 \,:\, S_n = x\}, \qquad x \geqslant 1.$$

Recall that the probability of eating at least x cookies before returning to the origin $\textcircled{0}$ is given by

$$\mathbb{P}\big(T_x^r < T_0^r \mid S_0 = 0\big) = \frac{1}{2}\prod_{l=2}^{x}\left(1 - \frac{2q}{l}\right), \qquad x \geqslant 1, \tag{5.9}$$

and that the random walk is recurrent, i.e. it returns to the origin $\textcircled{0}$ in finite time whenever $p < 1$, that means we have $\mathbb{P}\big(T_0^r < \infty \mid S_0 = 0\big) = 1$.

Proposition 5.6 *Let X denote the number of cookies eaten by the random walk before returning to the origin* $\textcircled{0}$*. We have*

$$\mathbb{P}(X = 0) = q, \qquad \mathbb{P}(X = 1) = p\widetilde{q}, \qquad \mathbb{P}(X = 2) = \frac{p\widetilde{p}q\widetilde{q}}{1 - pq},$$

and the distribution of X satisfies

$$\mathbb{P}(X = x) = pf(x) \prod_{l=1}^{x-1} (1 - f(l)), \quad x \geqslant 1, \tag{5.10}$$

where

$$f(l) := \frac{(q - p)\widetilde{q}}{(1 - (p/q)^{l+1})q^2} \in [0, 1], \qquad l \geqslant 1, \tag{5.11}$$

when $p \neq q$, *and*

$$f(l) := \frac{2\widetilde{q}}{l + 1} \in (0, 1], \qquad l \geqslant 1,$$

when $p = q = 1/2$.

Proof The probability $\mathbb{P}(X = 0)$ that the random walk eats no cookie before hitting the origin is the probability of going directly from $\textcircled{0}$ to $\textcircled{0}$ in one time step, which is q.

The probability $\mathbb{P}(X = 1)$ that the random walk eats exactly *one* cookie before hitting the origin is the probability of first moving from $\textcircled{0}$ to $\textcircled{1}$ in one time step and then back to $\textcircled{0}$ in one time step, that is $\widetilde{q} \times p$. When $p \neq q$, by Proposition 5.1 we have

$$\mathbb{P}(X = x) = \mathbb{P}\left(T_x^r < T_0^r \mid S_0 = 0\right) - \mathbb{P}\left(T_{x+1}^r < T_0^r \mid S_0 = 0\right)$$

$$= p \prod_{l=1}^{x-1} (1 - f(l)) - p \prod_{l=1}^{x} (1 - f(l))$$

$$= p \left(1 - (1 - f(x))\right) \prod_{l=1}^{x-1} (1 - f(l))$$

$$= pf(x) \prod_{l=1}^{x-1} (1 - f(l)).$$

$\square$

When $x = 2$, Proposition 5.6 yields

$$
\begin{aligned}
\mathbb{P}(X = 2) &= pf(2)(1 - f(1)) \\
&= \frac{(q-p)p\widetilde{q}}{(1-(p/q)^3)q^2}\left(1 - \frac{(q-p)\widetilde{q}}{(1-(p/q)^2)q^2}\right) \\
&= \frac{pq\widetilde{q}\,\widetilde{p}(q-p)}{q^3 - p^3} \\
&= \frac{pq\widetilde{q}\,\widetilde{p}}{q^2 + pq + p^2} \\
&= \frac{pq\widetilde{q}\,\widetilde{p}}{1 - pq} \\
&= pq\widetilde{q}\,\widetilde{p}\sum_{n=0}^{\infty}(pq)^n,
\end{aligned}
$$

which states that in order to eat two cookies, one has to take two steps up, two steps down, and to switch between states ① and ② for an arbitrary number of times n.

On the other hand, when $\widetilde{q} = 0$, the distribution of the number of cookies eaten by the random walk before returning to the origin ⓪ is given by

$$
\mathbb{P}(X = 0) = q, \quad \text{and} \quad \mathbb{P}(X = \infty) = 1 - q.
$$

The result of Proposition 5.6 can also be written as

$$
\mathbb{P}(X = x) = \frac{(q-p)p\widetilde{q}}{(1-(p/q)^{x+1})q^2}\prod_{l=2}^{x}\left(1 - \frac{(q-p)\widetilde{q}}{(1-(p/q)^l)q^2}\right), \quad x \geqslant 1, \qquad (5.12)
$$

when $p < 1/2$, and

$$
\mathbb{P}(X = x) = \frac{\widetilde{q}}{x+1}\prod_{l=2}^{x}\left(1 - \frac{2\widetilde{q}}{l}\right), \qquad x \geqslant 1, \qquad (5.13)
$$

if $p = q = 1/2$, with

$$
\sum_{x \geqslant 0}\mathbb{P}(X = x) = \frac{1}{2} + \sum_{x \geqslant 1}\frac{\widetilde{q}}{x+1}\prod_{l=2}^{x}\left(1 - \frac{2\widetilde{q}}{l}\right) = 1.
$$

Proposition 5.7 *Let $\widetilde{p} \in [0, 1)$. The number X of cookies eaten before returning to the origin ⓪ is finite with probability one, i.e. $\mathbb{P}(X < \infty) = 1$, if and only if and only if $p \leqslant 1/2$.*

Proof We may assume that $\tilde{q} \in (0, 1]$, otherwise the number of cookies eaten over time is clearly infinite. Using $f(l)$ defined in (5.11), we have

$$\mathbb{P}(X < \infty) = \sum_{x \geqslant 0} \mathbb{P}(X = x)$$

$$= q + p \sum_{x \geqslant 1} f(x) \prod_{l=1}^{x-1} (1 - f(l))$$

$$= q + p \sum_{x \geqslant 1} \left(\prod_{l=1}^{x-1} (1 - f(l)) - \prod_{l=1}^{x} (1 - f(l)) \right)$$

$$= q + p \lim_{n \to \infty} \sum_{x=1}^{n} \left(\prod_{l=1}^{x-1} (1 - f(l)) - \prod_{l=1}^{x} (1 - f(l)) \right)$$

$$= 1 - p \lim_{n \to \infty} \prod_{l=1}^{n} (1 - f(l))$$

$$= 1 - p \lim_{n \to \infty} \exp\left(\sum_{l=1}^{n} \log(1 - f(l)) \right) \tag{5.14}$$

$$\geqslant 1 - p \exp\left(- \lim_{n \to \infty} \sum_{l=1}^{n} f(l) \right). \tag{5.15}$$

(i) If $p < 1/2$, we have

$$\sum_{l \geqslant 1} f(l) = \frac{(q-p)\tilde{q}}{q^2} \sum_{l \geqslant 1} \frac{1}{1 - (p/q)^{l+1}} = +\infty,$$

hence $\mathbb{P}(X < \infty) = 1$ by (5.15).

(ii) If $p = q = 1/2$, we have

$$\sum_{l \geqslant 1} f(l) = \sum_{l \geqslant 1} \frac{2\tilde{q}}{l+1} = \infty,$$

hence by (5.15) we have $\mathbb{P}(X < \infty) = 1$ as well.

(iii) If $p \in (1/2, 1]$, we have

$$\sum_{l \geqslant 1} f(l) = \frac{(q-p)\tilde{q}}{q^2} \sum_{l \geqslant 1} \frac{1}{1 - (p/q)^{l+1}} < +\infty,$$

hence

$$- \infty < \sum_{l \geqslant 1} \log(1 - f(l)) \leqslant 0,$$

and the equality (5.14) shows that

$$\mathbb{P}(X < \infty) = 1 - p \exp\left(\lim_{n \to \infty} \sum_{l=1}^{n} \log(1 - f(l))\right) < 1.$$

□

When $\widetilde{q} = 0$ we have $\mathbb{P}(X < \infty) = \mathbb{P}(X = 0) = q$. From Remark 5.5 and the next proposition we note that in case $p = q = 1/2$ and $\widetilde{q} \in (1/2, 1)$ the mean number of eaten cookies $\mathbb{E}[X]$ is finite, while the mean return time $\mathbb{E}[T_0^r \mid S_0 = 0]$ is infinite.

Proposition 5.8 *Let* $\widetilde{p} \in [0, 1)$.

(i) *When* $p < 1/2$, *the average number* $\mathbb{E}[X]$ *of cookies eaten before returning to the origin* ⓪ *is finite, i.e.* $\mathbb{E}[X] < \infty$.

(ii) *In the critical case* $p = q = 1/2$, $\mathbb{E}[X]$ *is finite if and only if* $\widetilde{q} > 1/2$.

Proof

(i) Assume that $p < 1/2 < q$. We have

$$\mathbb{P}(X = x) = \frac{(q - p)p\widetilde{q}}{(1 - (p/q)^{x+1})q^2} \prod_{l=2}^{x}\left(1 + \frac{(p - q)\widetilde{q}}{(1 - (p/q)^l)q^2}\right)$$

$$\leqslant (q - p)p\widetilde{q}\frac{(1 + (p - q)\widetilde{q}/q^2)^{x-1}}{(1 - (p/q)^{x+1})q^2}, \quad x \geqslant 1,$$

hence

$$\mathbb{E}[X] = \sum_{x \geqslant 0} x\mathbb{P}(X = x)$$

$$\leqslant (q - p)\frac{p\widetilde{q}}{2q^2} \sum_{x \geqslant 1} x \frac{(1 + (p - q)\widetilde{q}/q^2)^{x-1}}{1 - (p/q)^{x+1}}$$

$$< \infty.$$

We note that we always have $1 + (p - q)\widetilde{q}/q^2 > 0$ since the equation

$$q^2 - 2q\widetilde{q} + \widetilde{q} = 0$$

has no real solution q, for any $\widetilde{q} \in (0, 1]$.

(ii) Assume that $p = q = 1/2$, and let $\varepsilon > 0$. Given $a_\varepsilon \geqslant 2$ such that

$$(1 + \varepsilon)z \leqslant \log(1 + z) \leqslant (1 - \varepsilon)z$$

for $z \in (-2\widetilde{q}/a_\varepsilon, 0)$ in a neighborhood of zero, we have

$$
\begin{aligned}
- 2(1 + \varepsilon)\widetilde{q} \log \frac{x}{a_\varepsilon} &= -2(1 + \varepsilon)\widetilde{q} \int_{a_\varepsilon}^{x} \frac{1}{y} dy \\
&\leqslant \int_{a_\varepsilon}^{x} \log\left(1 - \frac{2\widetilde{q}}{y}\right) dy \\
&\leqslant \sum_{l=a_\varepsilon+1}^{x} \log\left(1 - \frac{2\widetilde{q}}{l}\right) \\
&\leqslant \int_{a_\varepsilon+1}^{x+1} \log\left(1 - \frac{2\widetilde{q}}{y}\right) dy \\
&\leqslant -2(1 - \varepsilon)\widetilde{q} \int_{a_\varepsilon+1}^{x+1} \frac{1}{y} dy \\
&= -2(1 - \varepsilon)\widetilde{q} \log \frac{x + 1}{a_\varepsilon + 1},
\end{aligned}
$$

hence

$$\left(\frac{a_\varepsilon}{x}\right)^{2(1+\varepsilon)\widetilde{q}} \leqslant \prod_{l=a_\varepsilon+1}^{x} \left(1 - \frac{2\widetilde{q}}{l}\right) \leqslant \left(\frac{a_\varepsilon + 1}{x + 1}\right)^{2(1-\varepsilon)\widetilde{q}}, \qquad x \geqslant a_\varepsilon,$$

and

$$\sum_{x=1}^{a_\varepsilon} x \mathbb{P}(X = x) + \sum_{x>a_\varepsilon} x \left(\frac{a_\varepsilon}{x}\right)^{2(1+\varepsilon)\widetilde{q}} \leqslant \mathbb{E}[X]$$

$$\leqslant \sum_{x=1}^{a_\varepsilon} x \mathbb{P}(X = x) + \sum_{x>a_\varepsilon} x \left(\frac{a_\varepsilon + 1}{x + 1}\right)^{2(1-\varepsilon)\widetilde{q}},$$

hence $\mathbb{E}[X]$ is finite if $2(1 - \varepsilon)\widetilde{q} > 1$, and infinite if $2(1 + \varepsilon)\widetilde{q} < 1$. Since this statement is true for every $\varepsilon > 0$, we conclude that $\mathbb{E}[X]$ is finite if and only if $\widetilde{q} > 1/2$, and infinite if $\widetilde{q} < 1/2$.

In case $\widetilde{q} = 1/2$, by (5.13) we have

$$\mathbb{P}(X = x) = \frac{1/2}{x + 1} \prod_{l=2}^{x} \left(1 - \frac{1}{l}\right) = \frac{1}{2(x + 1)x}, \qquad x \geqslant 1,$$

Table 5.1 Behavior of the cookie random walk with $p = q = 1/2$ and $\tilde{p}, \tilde{q} \notin \{0, 1\}$

	$p = q$	
	$\tilde{p} < \tilde{q}$	$\tilde{q} \leqslant \tilde{p}$
Recurrence	Yes	Yes
Mean return time	Infinite	Infinite
Mean cookie count	Finite	Infinite

hence

$$\mathbb{E}[X] = \frac{1}{2} \sum_{x \geqslant 1} \frac{x}{x+1} \frac{1}{x} = +\infty.$$

$\square$

In case $p > 1/2$ or $\tilde{q} = 0$, we have $\mathbb{E}[X] = +\infty$ because $\mathbb{P}(X = \infty) > 0$. Table 5.1 summarizes some properties of the cookie random walk with $p = q = 1/2$ and $\tilde{p}, \tilde{q}$ different from 0 or 1.

5.5 Conditional Results

Lemma 5.9 *Assume that a cookie has just been eaten at state $x \geqslant 1$, after eating all cookies at states $1, 2, \ldots, x - 1$. Then, given that one hits $\boxed{x + 1}$ before hitting $\boxed{0}$, the probabilities of moving up to $\boxed{x + 1}$, resp. down to $\boxed{x - 1}$, are given by*

$$\mathbb{P}\big(S_1 = x + 1 \,\big|\, S_0 = \widehat{x} \text{ and } T^r_{x+1} < T^r_0\big) = \frac{(1 - (p/q)^{x+1})\tilde{p}q^2}{q^2(1 - (p/q)^{x+1}) + (p - q)\tilde{q}},$$

and

$$\mathbb{P}\big(S_1 = x - 1 \,\big|\, S_0 = \widehat{x} \text{ and } T^r_{x+1} < T^r_0\big) = \frac{(1 - (p/q)^{x-1})\tilde{q}p^2}{(1 - (p/q)^{x+1})q^2 + (p - q)\tilde{q}}$$

when $p \neq q$, and by

$$\mathbb{P}\big(S_1 = x + 1 \,\big|\, S_0 = \widehat{x} \text{ and } T^r_{x+1} < T^r_0\big) = \frac{\tilde{p}}{1 - 2\tilde{q}/(x + 1)}$$

and

$$\mathbb{P}\big(S_1 = x - 1 \,\big|\, S_0 = \widehat{x} \text{ and } T^r_{x+1} < T^r_0\big) = \frac{(x - 1)\tilde{q}/(x + 1)}{1 - 2\tilde{q}/(x + 1)}, \qquad x \geqslant 1,$$

when $p = q = 1/2$.

Proof We proceed similarly to the proof of Lemma 4.15.

(i) When $p \neq q$ we have

$$\mathbb{P}\big(S_1 = x + 1 \,\big|\, S_0 = \widehat{x} \text{ and } T^r_{x+1} < T^r_0\big)$$

$$= \widetilde{p}\,\frac{\mathbb{P}\big(T^r_{x+1} < T^r_0 \mid S_0 = x + 1\big)}{\mathbb{P}\big(T^r_{x+1} < T^r_0 \mid S_0 = \widehat{x}\big)}$$

$$= \frac{\widetilde{p}}{\mathbb{P}\big(T^r_{x+1} < T^r_0 \mid S_0 = \widehat{x}\big)},$$

as we have $\mathbb{P}\big(T^r_{x+1} < T^r_0 \mid S_0 = x + 1\big) = 1$. Next, we note that by (5.1) we have

$$\mathbb{P}\big(T^r_{x+1} < T^r_0 \mid S_0 = \widehat{x}\big) = 1 + \frac{(p - q)\widetilde{q}}{(1 - (p/q)^{x+1})q^2},$$

hence

$$\mathbb{P}(S_1 = x + 1 \mid S_0 = \widehat{x} \text{ and } T^r_{x+1} < T^r_0)$$

$$= \frac{\widetilde{p}}{1 + (p - q)\widetilde{q}/((1 - (p/q)^{x+1})q^2)}.$$

On the other hand, we have

$$\mathbb{P}(S_1 = x - 1 \mid S_0 = \widehat{x}, T^r_{x+1} < T^r_0) = \widetilde{q}\,\frac{\mathbb{P}\big(T^r_{x+1} < T^r_0 \mid S_0 = \widehat{x - 1}\big)}{\mathbb{P}\big(T^r_{x+1} < T^r_0 \mid S_0 = \widehat{x}\big)}$$

$$= \frac{(1 - (q/p)^{x-1})\widetilde{q}/(1 - (q/p)^{x+1})}{\mathbb{P}\big(T^r_{x+1} < T^r_0 \mid S_0 = \widehat{x}\big)},$$

and

$$\mathbb{P}\big(T^r_{x+1} < T^r_0 \mid S_0 = \widehat{x - 1}\big) = \frac{1 - (q/p)^{x-1}}{1 - (q/p)^{x+1}},$$

because the random walk evolves with probabilities (p, q) when started from state $\boxed{x - 1}$, hence we find

$$\mathbb{P}(S_1 = x - 1 \mid S_0 = \widehat{x} \text{ and } T^r_{x+1} < T^r_0)$$

$$= \widetilde{q}\,\frac{(1 - (q/p)^{x-1})/(1 - (q/p)^{x+1})}{1 + (p - q)\widetilde{q}/((1 - (p/q)^{x+1})q^2)}$$

$$= \frac{(1 - (p/q)^{x-1})\widetilde{q} p^2}{(1 - (p/q)^{x+1})q^2 + (p - q)\widetilde{q}}.$$

(ii) When $p = q = 1/2$ we note that, according to (5.2), $\mathbb{P}(T_{x+1}^r < T_0^r \mid S_0 = x)$
can be computed as

$$\mathbb{P}(T_{x+1}^r < T_0^r \mid S_0 = \widehat{x}) = \widetilde{p}\mathbb{P}(T_{x+1}^r < T_0^r \mid S_1 = x + 1)$$
$$+ \widetilde{q}\mathbb{P}(T_{x+1}^r < T_0^r \mid S_1 = x - 1)$$
$$= \widetilde{p} + \widetilde{q}\frac{x - 1}{x + 1},$$

hence

$$\mathbb{P}(S_1 = x + 1 \mid S_0 = x \text{ and } T_{x+1}^r < T_0^r)$$
$$= \frac{\widetilde{p}}{\widetilde{p} + (x - 1)\widetilde{q}/(x + 1)}$$
$$= \frac{\widetilde{p}}{1 - 2\widetilde{q}/(x + 1)}.$$

On the other hand, we have

$$\mathbb{P}(S_1 = x - 1 \mid S_0 = \widehat{x}, T_{x+1}^r < T_0^r) = \widetilde{q}\frac{\mathbb{P}(T_{x+1}^r < T_0^r \mid S_0 = \widehat{x - 1})}{\mathbb{P}(T_{x+1}^r < T_0^r \mid S_0 = \widehat{x})}$$
$$= \frac{(x - 1)\widetilde{q}/(x + 1)}{\mathbb{P}(T_{x+1}^r < T_0^r \mid S_0 = \widehat{x})},$$

and

$$\mathbb{P}(T_{x+1}^r < T_0^r \mid S_0 = \widehat{x - 1}) = \frac{x - 1}{x + 1},$$

because the random walk becomes symmetric when started from state $\boxed{x - 1}$.
Hence, we find

$$\mathbb{P}(S_1 = x - 1 \mid S_0 = \widehat{x} \text{ and } T_{x+1}^r < T_0^r) = \frac{(x - 1)\widetilde{q}/(x + 1)}{\widetilde{p} + (x - 1)\widetilde{q}/(x + 1)}$$
$$= \frac{(x - 1)\widetilde{q}/(x + 1)}{1 - 2\widetilde{q}/(x + 1)}.$$

$\square$

Proposition 5.10 *Assume that $p = q = 1/2$. The mean time to reach state $\textcircled{x}$ from a cookie at state $\textcircled{1}$ given that one* does *not hit $\textcircled{0}$ is given for $x \geqslant 2$ by*

$$\mathbb{E}[T_x^r \mid S_0 = \hat{1} \text{ and } T_x^r < T_0^r]$$

$$= x - 1 + \frac{4\tilde{q}}{3}\left(\frac{x(x-1)}{2} - 2(x-1)\tilde{p} + 2(\tilde{p} - \tilde{q})\tilde{p}\sum_{k=1}^{x-1}\frac{1}{k+1-2\tilde{q}}\right).$$

Proof Since $p = q = 1/2$, Proposition 4.17 shows that

$$\mathbb{E}[T_{x+1}^r \mid S_0 = x - 1, \ T_{x+1}^r < T_0^r] = \frac{(x+1)^2 - (x-1)^2}{3} = \frac{4x}{3}, \qquad x \geqslant 2,$$

while for $x = 1$ we have $\mathbb{E}[T_2^r \mid S_0 = 0, \ T_2^r < T_0^r] = 2$, and $\mathbb{P}(S_1 = 0 \mid S_0 = \hat{1} \text{ and } T_2^r < T_0^r) = 0$. Hence, given that a cookie has just been eaten at state $x \geqslant 1$ after eating all cookies at states $1, 2, \ldots, x - 1$, the mean time to reach the next cookie at state $\boxed{x+1}$ given that one *does not hit $\textcircled{0}$* is given from Lemma 5.9 as

$$\mathbb{E}[T_{x+1}^r \mid S_0 = \hat{x}, \ T_{x+1}^r < T_0^r] = \mathbb{P}(S_1 = x + 1 \mid S_0 = \hat{x} \text{ and } T_{x+1}^r < T_0^r)$$

$$+ \mathbb{P}(S_1 = x - 1 \mid S_0 = \hat{x} \text{ and } T_{x+1}^r < T_0^r)$$

$$\times (1 + \mathbb{E}[T_{x+1}^r \mid S_0 = x - 1, \ T_{x+1}^r < T_0^r])$$

$$= \mathbb{P}(S_1 = x + 1 \mid S_0 = \hat{x} \text{ and } T_{x+1}^r < T_0^r)$$

$$+ \mathbb{P}(S_1 = x - 1 \mid S_0 = \hat{x} \text{ and } T_{x+1}^r < T_0^r)\left(1 + \frac{(x+1)^2 - (x-1)^2}{3}\right)$$

$$= \mathbb{P}(S_1 = x + 1 \mid S_0 = \hat{x} \text{ and } T_{x+1}^r < T_0^r)$$

$$+ \mathbb{P}(S_1 = x - 1 \mid S_0 = \hat{x} \text{ and } T_{x+1}^r < T_0^r)\left(1 + \frac{(x+1)^2 - (x-1)^2}{3}\right)$$

$$= 1 + \frac{(x-1)\tilde{q}/(x+1)}{1 - 2\tilde{q}/(x+1)} \times \frac{(x+1)^2 - (x-1)^2}{3}$$

$$= 1 + \frac{4\tilde{q}x(x-1)/(x+1)}{3(1 - 2\tilde{q}/(x+1))}$$

$$= 1 + \frac{((x+1)^2 - (x-1)^2)}{3}\frac{(x-1)\tilde{q}/(x+1)}{1 - 2\tilde{q}/(x+1)}$$

$$= 1 + \frac{4\tilde{q}x(x-1)/(x+1)}{3(1 - 2\tilde{q}/(x+1))}$$

$$= 1 + \frac{4\tilde{q}x(x-1)}{3(x+1-2\tilde{q})}$$

$$= 1 + \frac{4x}{3} \times \frac{\widetilde{q} - 2\widetilde{q}/(x+1)}{1 - 2\widetilde{q}/(x+1)}$$

$$= 1 + \frac{4\widetilde{q}}{3} \times \frac{x(x-1)}{x+1-2\widetilde{q}}, \qquad x \geqslant 1,$$

which yields $1 + (2x - 2)/3$ when $\widetilde{p} = \widetilde{q} = 1/2$. Hence for $x \geqslant 2$ we have

$$\mathbb{E}[T_x^r \mid S_0 = \hat{1}, \ T_x^r < T_0^r] = \sum_{k=1}^{x-1} \mathbb{E}[T_{k+1}^r \mid S_0 = \widehat{k}, \ T_{k+1}^r < T_0^r]$$

$$= x - 1 + \frac{4\widetilde{q}}{3} \sum_{k=1}^{x-1} \frac{k(k-1)}{k+1-2\widetilde{q}}$$

$$= x - 1 + \frac{4\widetilde{q}}{3} \sum_{k=1}^{x-1} k - \frac{8\widetilde{p}\widetilde{q}}{3} \sum_{k=1}^{x-1} \frac{k}{k+1-2\widetilde{q}}$$

$$= x - 1 + \frac{2\widetilde{q}x(x-1)}{3} - \frac{8\widetilde{p}\widetilde{q}}{3} \sum_{k=1}^{x-1} \frac{k}{k+1-2\widetilde{q}}$$

$$= x - 1 + \frac{2\widetilde{q}x(x-1)}{3} - \frac{8\widetilde{p}\widetilde{q}}{3}(x-1) + \frac{8\widetilde{p}\widetilde{q}}{3}(\widetilde{p} - \widetilde{q}) \sum_{k=1}^{x-1} \frac{1}{k+1-2\widetilde{q}}$$

$$= x - 1 + \frac{4\widetilde{q}}{3} \left(\frac{x(x-1)}{2} - 2(x-1)\widetilde{p} + 2(\widetilde{p} - \widetilde{q})\widetilde{p} \sum_{k=1}^{x-1} \frac{1}{k+1-2\widetilde{q}} \right).$$

$\square$

We note that for $x \geqslant 1$ we have

$$\mathbb{P}(S_1 = 1 \mid S_0 = 0 \text{ and } T_x^r < T_0^r) = 1 \quad \text{and} \quad \mathbb{P}(S_1 = 0 \mid S_0 = 0 \text{ and } T_x^r < T_0^r) = 0,$$

hence by Proposition 5.10 we have

$$\mathbb{E}[T_x^r \mid S_0 = 0 \text{ and } T_x^r < T_0^r]$$

$$= \mathbb{P}(S_1 = 1 \mid S_0 = 0 \text{ and } T_x^r < T_0^r)\big(1 + \mathbb{E}[T_x^r \mid S_0 = \hat{1} \text{ and } T_x^r < T_0^r]\big)$$

$$\quad + \mathbb{P}(S_1 = 0 \mid S_0 = 0 \text{ and } T_x^r < T_0^r)$$

$$= 1 + \mathbb{E}[T_x^r \mid S_0 = \hat{1} \text{ and } T_x^r < T_0^r]$$

$$= x + \frac{4\widetilde{q}}{3} \left(\frac{x(x-1)}{2} - 2(x-1)\widetilde{p} + 2(\widetilde{p} - \widetilde{q})\widetilde{p} \sum_{k=1}^{x-1} \frac{1}{k+1-2\widetilde{q}} \right).$$

When $p = q = \widetilde{p} = \widetilde{q} = 1/2$ we recover the classical expression

$$\mathbb{E}[T_x^r \mid S_0 = 1, \ T_x^r < T_0^r] = x - 1 + \frac{2}{3} \sum_{k=1}^{x-2} k$$

$$= x - 1 + \frac{(x-1)(x-2)}{3}$$

$$= \frac{x^2 - 1}{3}, \qquad x \geqslant 2,$$

cf. Proposition 4.17. The mean time $\mathbb{E}[T_x^r \mid S_0 = \widehat{1}$ and $T_x^r < T_0^r]$ to reach state $\widehat{x}$ from state $\widehat{1}$ given that one *does not hit* $\widehat{0}$ can similarly be computed from Proposition 4.17 and Lemma 5.9 when $p \neq q$. Indeed, when $p \neq q$, Proposition 4.17 shows that

$$\mathbb{E}[T_{x+1}^r \mid S_0 = x - 1, \ T_{x+1}^r < T_0^r]$$

$$= \frac{(x+1)(1 + (q/p)^{x+1})}{(p-q)(1 - (q/p)^{x+1})} - \frac{(x-1)(1 + (q/p)^{x-1})}{(p-q)(1 - (q/p)^{x-1})}$$

$$= \frac{(x+1)(1+(q/p)^{x+1})(1-(q/p)^{x-1}) - (x-1)(1+(q/p)^{x-1})(1-(q/p)^{x+1})}{(p-q)(1 - (q/p)^{x+1})(1 - (q/p)^{x-1})}$$

$$= 2\frac{(p-q)/p^2 + x(q/p)^{x+1} - x(q/p)^{x-1}}{(p-q)((p-q)/q^2 - (q/p)^{x-1} - (q/p)^{x+1})}, \qquad x \geqslant 2.$$

Hence from Lemma 5.9 we can similarly compute the mean time to reach the next cookie at state $\boxed{x+1}$ given that a cookie has just been eaten at state $x \geqslant 1$ and *one does not hit* $\widehat{0}$, after eating all cookies at states $1, 2, \ldots, x - 1$, as

$$\mathbb{E}[T_{x+1}^r \mid S_0 = \widehat{x}, \ T_{x+1}^r < T_0^r]$$

$$= \mathbb{P}(S_1 = x + 1 \mid S_0 = \widehat{x} \text{ and } T_{x+1}^r < T_0^r)$$

$$+ \mathbb{P}(S_1 = x - 1 \mid S_0 = \widehat{x} \text{ and } T_{x+1}^r < T_0^r)$$

$$\times (1 + \mathbb{E}[T_{x+1}^r \mid S_0 = x - 1, \ T_{x+1}^r < T_0^r]),$$

$x \geqslant 1$, and

$$\mathbb{E}[T_x^r \mid S_0 = 1, \ T_x^r < T_0^r] = \sum_{k=1}^{x-1} \mathbb{E}[T_{k+1}^r \mid S_0 = k, T_{k+1}^r < T_0^r], \qquad x \geqslant 2.$$

Notes

See e.g. Benjamini and Wilson (2003) and Antal and Redner (2005) for further reading on excited random walks.

Exercises

Exercise 5.1 (Antal and Redner (2005), § 5)
Consider a cookie-excited random walk $(S_n)_{n \geqslant 0}$ on the half line $\mathbb{Z}_+$, with probabilities $(p, q) = (1/2, 1/2)$ of moving up and down without cookies, and probabilities $(\widetilde{p}, \widetilde{q})$ of moving up and down on cookie locations, with $\widetilde{p} > \widetilde{q}$.

We assume that

- $(S_n)_{n \geqslant 0}$ starts at $S_0 = 0$ with no cookie at state $\textcircled{0}$,
- every cookie location at states $\textcircled{i}$, $i \geqslant 1$, contains initially a same number $k \geqslant 1$ of cookies, and
- only a single cookie can be eaten at each step.

(a) Give the number of cookies initially contained in the region $\{1, 2, \ldots, L\}$, $L \geqslant 1$.
(b) Give the minimum number of time steps needed to consume all cookies by traveling within $\{1, 2, \ldots, L\}$.
(c) Assuming a positive average drift $\widetilde{p} - \widetilde{q} > 0$ on cookie locations at every time step, give the average number of time steps needed to travel from state $\textcircled{1}$ to state $\textcircled{L}$, assuming that all states contain cookies.
(d) Find a condition on $\widetilde{p}$ and k ensuring the consumption of all cookies while traveling from from $\textcircled{1}$ to $\textcircled{L}$.
(e) Find a sufficient condition based on $\widetilde{p}$ and k for the transience of this cookie random walk.

Exercise 5.2 (Antal and Redner (2005)) .A random walk $(S_n)_{n \geqslant 0}$ with cookies on $\{1, 2, 3, \ldots\}$ is symmetric in the absence of cookies, and restarts with probabilities p and $q = 1 - p$ of moving up, resp. down, when it encounters a cookie, where $p \in [0, 1)$. The random walk starts at state $\textcircled{0}$, which is empty of cookie.

For any $x \geqslant 1$, let τ_x denote the first hitting time

$$\tau_x := \inf\{n \geqslant 1 : S_n = x\}, \qquad x \geqslant 1.$$

Recall that the probability of eating at least x cookies before returning to the origin $\textcircled{0}$ is given by

$$\mathbb{P}(\tau_x < \tau_0 \mid S_0 = 0) = \frac{1}{2} \prod_{l=2}^{x} \left(1 - \frac{2q}{l} \right), \qquad x \geqslant 1, \tag{5.16}$$

and that the random walk is recurrent, i.e. it returns to the origin $\textcircled{0}$ in finite time whenever $p < 1$, that means we have $\mathbb{P}(\tau_0 < \infty \mid S_0 = 0) = 1$.

(a) Let X denote the number of cookies eaten by the random walk before returning to the origin $\textcircled{0}$. Show that

$$\mathbb{P}(X = 0) = 1/2, \qquad \mathbb{P}(X = 1) = q/2,$$

and, using (5.16), that the distribution of satisfies

$$\mathbb{P}(X = x) = \frac{q}{x+1} \prod_{l=2}^{x} \left(1 - \frac{2q}{l} \right), \qquad x \geqslant 2. \tag{5.17}$$

(b) Show from (5.17) that the average number $\mathbb{E}[X]$ of cookies eaten before returning to the origin $\textcircled{0}$ is finite, i.e. $\mathbb{E}[X] < \infty$, if and only if $q > 1/2$. *Hint:* There exist constants $c_q, C_q > 0$ such that

$$\frac{c_q}{x^{2q}} \leqslant \prod_{l=2}^{x} \left(1 - \frac{2q}{l} \right) \leqslant \frac{C_q}{x^{2q}}, \qquad x \geqslant 2.$$

Chapter 6
Convergence to Equilibrium

This chapter is concerned with the large time behavior of Markov chains, including the computation of their limiting and stationary distributions. Here the notions of recurrence, transience, and classification of states introduced in the previous chapter play a major role. We also derive quantitative bounds for the convergence of a Markov chain to its stationary distribution. The Markov Chain Monte Carlo (MCMC) method presented in Sect. 6.2 is widely used for statistical estimation based on the Markov property.

6.1 Limiting and Stationary Distributions

This section gathers some basic facts on the long run behavior of Markov chains, characterized by their limiting and stationary distributions. It is generally assumed that the state space S is countable and possibly infinite, while finite state spaces are treated as particular cases.

Limiting Distributions

Definition 6.1 A Markov chain $(X_n)_{n \geqslant 0}$ is said to admit a *limiting probability distribution* if the following conditions are satisfied:

(i) the limits

$$\lim_{n \to \infty} \mathbb{P}(X_n = j \mid X_0 = i) \tag{6.1}$$

© The Author(s), under exclusive license to Springer Nature Switzerland AG 2024
N. Privault, *Discrete Stochastic Processes*, Springer Undergraduate
Mathematics Series, https://doi.org/10.1007/978-3-031-65820-4_6

exist for all $i, j \in S$, and

(ii) they form a *probability distribution* on S, i.e.

$$\sum_{j \in S} \lim_{n \to \infty} \mathbb{P}(X_n = j \mid X_0 = i) = 1, \tag{6.2}$$

for all $i \in S$.

Note that Condition (6.2) is always satisfied if the limits (6.1) exist and the state space S is finite. As an example, consider the two-state Markov chain, whose transition matrix has the form

$$P = \begin{bmatrix} 1-a & a \\ b & 1-b \end{bmatrix}, \tag{6.3}$$

with $a \in [0, 1]$ and $b \in [0, 1]$.

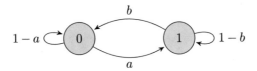

The matrix power

$$P^n = \begin{bmatrix} 1-a & a \\ b & 1-b \end{bmatrix}^n = \underbrace{\begin{bmatrix} 1-a & a \\ b & 1-b \end{bmatrix} \times \cdots \times \begin{bmatrix} 1-a & a \\ b & 1-b \end{bmatrix}}_{n \text{ times}}$$

of the transition matrix P can be computed for all $n \geq 0$ as

$$P^n = \frac{1}{a+b} \begin{bmatrix} b + a(1-a-b)^n & a(1-(1-a-b)^n) \\ b(1-(1-a-b)^n) & a + b(1-a-b)^n \end{bmatrix}, \quad n \geq 0,$$

which can be obtained in Mathematica via the command

MatrixPower[1-a,a,b,1-b,n].

The two-state Markov chain has a limiting distribution $[\pi_0, \pi_1]$ independent of the initial state, and given by

$$\lim_{n \to \infty} P^n = \begin{bmatrix} \dfrac{b}{a+b} & \dfrac{a}{a+b} \\[2ex] \dfrac{b}{a+b} & \dfrac{a}{a+b} \end{bmatrix},$$

i.e.

$$[\pi_0, \pi_1] = \left[\frac{b}{a+b}, \frac{a}{a+b} \right], \tag{6.4}$$

provided that $(a, b) \neq (0, 0)$ and $(a, b) \neq (1, 1)$, while the corresponding mean return times are given by

$$(\mu_0(0), \mu_1(1)) = \left(1 + \frac{a}{b}, 1 + \frac{b}{a} \right),$$

see e.g. Relation (5.3.3) in Privault (2018), i.e. the limiting probabilities are given by the mean return time inverses, as

$$[\pi_0, \pi_1] = \left[\frac{b}{a+b}, \frac{a}{a+b} \right] = \left[\frac{1}{\mu_0(0)}, \frac{1}{\mu_1(1)} \right]$$
$$= \left[\frac{\mu_1(0)}{\mu_0(1) + \mu_1(0)}, \frac{\mu_0(1)}{\mu_0(1) + \mu_1(0)} \right].$$

Theorem 6.2 (Karlin and Taylor (1998), Theorem IV.4.1) *Consider a Markov chain $(X_n)_{n \geq 0}$ satisfying the following three conditions:*

  (i) irreducibility,
 (ii) recurrence, and
(iii) aperiodicity.

Then, the chain $(X_n)_{n \geq 0}$ admits the limiting distribution

$$\lim_{n \to \infty} \mathbb{P}(X_n = j \mid X_0 = i) = \frac{1}{\mu_j(j)}, \qquad i, j \in S, \tag{6.5}$$

independently of the initial state $i \in S$, where

$$\mu_j(j) = \mathbb{E}[T_j^r \mid X_0 = j] \in [1, \infty]$$

is the mean return time to state $\textcircled{j} \in S$.

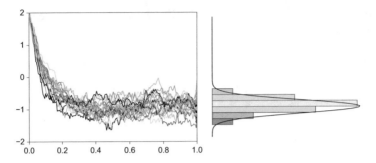

Fig. 6.1 Convergence in distribution

In Theorem 6.2, Condition (i), resp. Condition (ii), is satisfied from Proposition 1.23, resp. from Proposition 1.13, provided that at least one state is aperiodic, resp. recurrent, since the chain is irreducible.

Example

Figure 6.1 illustrates the convergence in distribution of the Markov chain $(Y_n)_{n \geqslant 0}$ when it is not started from its stationary distribution, and can be plotted using the ®code_12_page_154.R.

Stationary Distributions

In what follows, we let $\mathcal{P}_N$ denote the set of probability distributions on $\{1, \ldots, N\}$, which are represented by vectors $\mu = (\mu_i)_{i=1,\ldots,N}$ in $[0, 1]$ such that

$$\sum_{i=1}^{N} \mu_i = 1.$$

Definition 6.3 A *probability distribution* $\pi = (\pi_i)_{i \in S}$ on S is said to be *stationary* if, starting X_0 at time 0 with the distribution $(\pi_i)_{i \in S}$, it turns out that the distribution of X_1 is still $(\pi_i)_{i \in S}$ at time 1.

In other words, $(\pi_i)_{i \in S}$ is stationary for the Markov chain with transition matrix P if, letting

$$\mathbb{P}(X_0 = i) := \pi_i, \qquad i \in S,$$

at time 0, implies

$$\mathbb{P}(X_1 = i) = \mathbb{P}(X_0 = i) = \pi_i, \quad i \in S,$$

at time 1. This also means that

$$\pi_j = \mathbb{P}(X_1 = j) = \sum_{i \in S} \mathbb{P}(X_1 = j \mid X_0 = i)\mathbb{P}(X_0 = i) = \sum_{i \in S} \pi_i P_{i,j}, \quad j \in S,$$

i.e. the distribution π is stationary if and only if the vector π is *invariant* (or stationary) by the matrix P, that means

$$\pi = \pi P. \tag{6.6}$$

Example

Figure 6.2 considers the Markov chain $(Y_n)_{n \geqslant 0}$ recursively defined as

$$Y_{n+1} = Y_n + b + aY_n + \sigma Z_n,$$

which admits the $\mathcal{N}\big(-b/a, \sqrt{\sigma^2/2/(-a)}\big)$ Gaussian distribution as stationary distribution, where $(Z_n)_{n \geqslant 1}$ is a sequence of $\mathcal{N}(0, 1)$ centered Gaussian random variables. We note that the process $(Y_n)_{n \geqslant 0}$ remains in the $\mathcal{N}\big(-b/a, \sqrt{\sigma^2/2/(-a)}\big)$ Gaussian distribution if Y_0 is started from this distribution. Figure 6.2 can be plotted using the ®R code_13_page_155.R.

More generally, assuming that X_n has the invariant (or stationary) distribution π at time n, i.e. $\mathbb{P}(X_n = i) = \pi_i, i \in S$, we have

$$\mathbb{P}(X_{n+1} = j) = \sum_{i \in S} \mathbb{P}(X_{n+1} = j \mid X_n = i)\mathbb{P}(X_n = i)$$

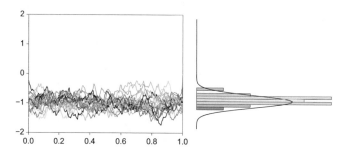

Fig. 6.2 Stationarity in distribution

$$= \sum_{i \in S} P_{i,j} \mathbb{P}(X_n = i) = \sum_{i \in S} P_{i,j} \pi_i$$

$$= [\pi P]_j = \pi_j, \qquad j \in S,$$

since the Markov chain $(X_n)_{n \geqslant 0}$ is *time homogeneous*, i.e. its transition matrix P remains constant over time, hence

$$\mathbb{P}(X_n = j) = \pi_j, \quad j \in S, \quad \Longrightarrow \quad \mathbb{P}(X_{n+1} = j) = \pi_j, \quad j \in S.$$

By induction on $n \geqslant 0$, this yields

$$\mathbb{P}(X_n = j) = \pi_j, \qquad j \in S, \quad n \geqslant 1,$$

i.e. the chain $(X_n)_{n \geqslant 0}$ remains in the same distribution π at all times $n \geqslant 1$, provided that it has been started with the stationary distribution π at time $n = 0$.

Relation (6.6) can be rewritten as the *global balance condition*

$$\sum_{i \in S} \pi_i P_{i,k} = \pi_k = \pi_k \sum_{j \in S} P_{k,j} = \sum_{j \in S} \pi_k P_{k,j}, \qquad (6.7)$$

which is illustrated in Fig. 6.3.

On the other hand, the $(X_n)_{n \geqslant 0}$ is said to satisfy the *detailed balance* (or *reversibility*) condition with respect to the probability distribution $\pi = (\pi_i)_{i \in S}$ if

$$\pi_i P_{i,j} = \pi_j P_{j,i}, \qquad i, j \in S, \qquad (6.8)$$

see Fig. 6.4.

Lemma 6.4 *The detailed balance condition (6.8) implies the global balance condition (6.7).*

Fig. 6.3 Global balance condition

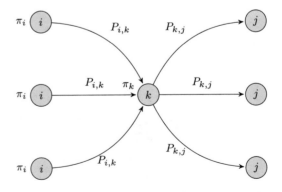

Fig. 6.4 Detailed balance
condition (discrete time)

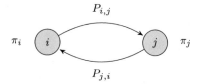

Proof By summation over $i \in S$ in (6.8) we have

$$\sum_{i \in S} \pi_i P_{i,j} = \sum_{i \in S} \pi_j P_{j,i} = \pi_j \sum_{i \in S} P_{j,i} = \pi_j, \quad j \in S,$$

which shows that $\pi P = \pi$, i.e. π is a stationary distribution for P. $\qquad\square$

The next result shows that existence of a limiting distribution implies the existence of a stationary distribution when the chain $(X_n)_{n \geqslant 0}$ has a finite state space.

Proposition 6.5 *Assume that* $S = \{0, 1, \dots, N\}$ *is* finite *and that the limits*

$$\pi_j := \lim_{n \to \infty} \mathbb{P}(X_n = j \mid X_0 = i) = \lim_{n \to \infty} [P^n]_{i,j}$$

exist for all $j \in S$ *and are independent of the initial state* $i \in S$, *i.e. we have*

$$\lim_{n \to \infty} P^n = \begin{bmatrix} \pi_0 & \pi_1 & \cdots & \pi_N \\ \pi_0 & \pi_1 & \cdots & \pi_N \\ \vdots & \vdots & \ddots & \vdots \\ \pi_0 & \pi_1 & \cdots & \pi_N \end{bmatrix}.$$

Then for every $i = 0, 1, \dots, N$, *the vector* $\pi := (\pi_j)_{j \in \{0,1,\dots,N\}}$ *is a stationary distribution and we have*

$$\pi = \pi P, \tag{6.9}$$

i.e. π *is invariant (or stationary) by* P.

Proposition 6.5 can be applied in particular when the limiting distribution $\pi_j := \lim_{n \to \infty} \mathbb{P}(X_n = j \mid X_0 = i)$ does not depend on the initial state $\textcircled{i}$, i.e.

$$\lim_{n \to \infty} P^n = \begin{bmatrix} \pi_0 & \pi_1 & \cdots & \pi_N \\ \pi_0 & \pi_1 & \cdots & \pi_N \\ \vdots & \vdots & \ddots & \vdots \\ \pi_0 & \pi_1 & \cdots & \pi_N \end{bmatrix}.$$

For example, the limiting distribution (6.4) of the two-state Markov chain is also an invariant distribution, i.e. it satisfies (6.6). In particular we have the following result.

Theorem 6.6 (Karlin and Taylor (1998), Theorem IV.4.2) *Assume that the Markov chain $(X_n)_{n \geqslant 0}$ satisfies the following three conditions:*

(i) *irreducibility,*
(ii) *positive recurrence, and*
(iii) *aperiodicity.*

Then the chain $(X_n)_{n \geqslant 0}$ admits the limiting distribution

$$\pi_j := \lim_{n \to \infty} \mathbb{P}(X_n = j \mid X_0 = i) = \lim_{n \to \infty} [P^n]_{i,j} = \frac{1}{\mu_j(j)}, \qquad i, j \in S,$$

independently of the initial state $i \in S$, which also forms a stationary distribution $(\pi_j)_{j \in S} = (1/\mu_j(j))_{j \in S}$, uniquely determined by the equation

$$\pi = \pi P.$$

In Theorem 6.6 above, Condition (ii), is satisfied from Proposition 1.23, provided that at least one state is aperiodic, since the chain is irreducible. See also pages 170–171 in Privault (2018) for counterexamples.

In view of Theorem 1.20, we have the following corollary of Theorem 6.6:

Corollary 6.7 *Consider an irreducible aperiodic Markov chain with finite state space. Then, the limiting probabilities*

$$\pi_i := \lim_{n \to \infty} \mathbb{P}(X_n = i \mid X_0 = j) = \frac{1}{\mu_i(i)}, \qquad i, j \in S,$$

exist and form a stationary distribution which is uniquely determined by the equation

$$\pi = \pi P.$$

Corollary 6.7 can also be applied separately to derive a stationary distribution on each closed component of a reducible chain.

The following theorem gives sufficient conditions for the existence of a stationary distribution, without requiring aperiodicity or finiteness of the state space. Note that the limiting distribution may not exist in this case, as can be checked for the two-state chain (6.3) with $a = b = 1$. See also Problem 6.9 and Exercise 7.21 in Privault (2018) for an example of a null recurrent chain which does not admit a stationary distribution.

Theorem 6.8 (Bosq and Nguyen (1996), Theorem 4.1) *Consider a Markov chain* $(X_n)_{n \geqslant 0}$ *satisfying the following two conditions:*

(i) irreducibility, and
(ii) positive recurrence.

Then, the probabilities

$$\pi_i = \frac{1}{\mu_i(i)}, \qquad i \in S,$$

form a stationary distribution which is uniquely determined by the equation $\pi = \pi P$.

Note that the conditions stated in Theorem 6.8 are sufficient, but they are not all necessary. For example, Condition (ii) is not necessary as the trivial constant chain, whose transition matrix $P = I$ is reducible, does admit a stationary distribution. Note that the positive recurrence assumption in Theorem 6.2 is required in general on infinite state spaces.

As a consequence of Corollary 1.21 we have the following corollary of Theorem 6.8, which does not require aperiodicity for the stationary distribution to exist.

Corollary 6.9 *Let* $(X_n)_{n \geqslant 0}$ *be an irreducible Markov chain with* finite *state space* S. *Then, the probabilities*

$$\pi_k = \frac{1}{\mu_k(k)}, \qquad k \in S,$$

form a stationary distribution which is uniquely determined by the equation

$$\pi = \pi P.$$

6.2   Markov Chain Monte Carlo: MCMC

Generating Random Samples from a Target Distribution

The Markov Chain Monte Carlo (MCMC) method, or Metropolis algorithm, can be used to generate random samples according to a target distribution $\pi = (\pi_i)_{i \in S}$ via a Markov chain that admits π as a limiting and stationary distribution. It can be applied in particular in the setting of large state spaces S, cf. e.g. Chap. 7. See Diaconis (2009) for a review of applications including a cryptography example, the analysis of algorithms and their complexity in computer science, and particle filters for tracking and filtering.

If the transition matrix P satisfies the detailed balance condition (6.8) with respect to π, then the probability distribution of X_n will naturally converge to the stationary distribution π in the long run, e.g. under the hypotheses of Theorem 6.6, i.e. when the chain $(X_k)_{k\in\mathbb{N}}$ is positive recurrent, aperiodic, and irreducible.

In general, however, π and P may not satisfy by the global or detailed balance conditions (6.7) or (6.8). In this case, starting from a proposal matrix P, one can construct a modified transition matrix $\widetilde{P}$ that will satisfy the detailed balance condition with respect to π. This modified transition matrix $\widetilde{P}$ is defined by

$$
\begin{aligned}
\widetilde{P}_{i,j} &:= \min\left(P_{i,j}, \frac{\pi_j}{\pi_i} P_{j,i} \right) \\[2mm]
&= P_{i,j} \times \min\left(1, \frac{\pi_j P_{j,i}}{\pi_i P_{i,j}} \right) \\[2mm]
&= \begin{cases} \dfrac{\pi_j}{\pi_i} P_{j,i} & \text{if } \pi_j P_{j,i} \leqslant \pi_i P_{i,j}, \\[4mm] P_{i,j} & \text{if } \pi_j P_{j,i} \geqslant \pi_i P_{i,j}, \end{cases}
\end{aligned}
\tag{6.10}
$$

for $i \neq j$. We note that

$$
\sum_{\substack{j\in S \\ j\neq i}} \widetilde{P}_{i,j} \leqslant \sum_{\substack{j\in S \\ j\neq i}} P_{i,j} \leqslant 1,
$$

and for $i \in S$ we let

$$
\begin{aligned}
\widetilde{P}_{i,i} &:= 1 - \sum_{\substack{j\in S \\ j\neq i}} \widetilde{P}_{i,j} \\[2mm]
&= P_{i,i} + \sum_{\substack{j\in S \\ j\neq i}} P_{i,j} \left(1 - \min\left(1, \frac{\pi_j P_{j,i}}{\pi_i P_{i,j}} \right) \right) \\[2mm]
&= P_{i,i} + \sum_{\substack{j\in S \\ j\neq i}} P_{i,j} \left(1 - \frac{\pi_j P_{j,i}}{\pi_i P_{i,j}} \right)^{+} \\[2mm]
&= P_{i,i} + \sum_{\substack{j\in S \\ j\neq i}} \left(P_{i,j} - \frac{\pi_j P_{j,i}}{\pi_i} \right)^{+}.
\end{aligned}
$$

Clearly, we have $\widetilde{P} = P$ when the detailed balance (or reversibility) condition (6.8) is satisfied by P. In the general case, we can check that for $i \neq j$, we have

$$
\pi_i \widetilde{P}_{i,j} = \left\{
\begin{array}{l}
P_{j,i} \pi_j = \pi_j \widetilde{P}_{j,i} \ \text{ if } \ \pi_j P_{j,i} \leqslant \pi_i P_{i,j}, \\[2mm]
\pi_i P_{i,j} = \pi_j \widetilde{P}_{j,i} \ \text{ if } \ \pi_j P_{j,i} \geqslant \pi_i P_{i,j},
\end{array}
\right\} = \pi_j \widetilde{P}_{j,i},
$$

hence $\widetilde{P}$ satisfies the detailed balance condition with respect to π (the condition is obviously satisfied when $i = j$). Therefore, the random simulation of $(\widetilde{X}_n)_{n \geqslant 0}$ according to the transition matrix $\widetilde{P}$ will provide samples of the distribution π in the long run as n tends to infinity, provided that the chain $(\widetilde{X}_n)_{n \geqslant 0}$ is positive recurrent, aperiodic, and irreducible.

In standard MCMC sampling we make the following assumption, which is typically satisfied by taking $P_{i,j} := \varphi(i - j)$ with φ a Gaussian type density kernel.

Assumption (B) *The transition matrix P is symmetric i.e.* $P_{i,j} = P_{j,i} > 0$, $i, j \in \mathsf{S}$.

Under Assumption (B), the modified transition matrix $\widetilde{P}$ simplifies to

$$
\widetilde{P}_{i,j} := P_{i,j} \times \min\left(1, \frac{\pi_j}{\pi_i}\right) = \min\left(P_{i,j}, \frac{\pi_j}{\pi_i} P_{i,j}\right) = \left\{
\begin{array}{ll}
P_{i,j} \dfrac{\pi_j}{\pi_i} & \text{if } \pi_j \leqslant \pi_i, \\[4mm]
P_{i,j} & \text{if } \pi_j \geqslant \pi_i,
\end{array}
\right.
$$

for $i \neq j$, with

$$
\widetilde{P}_{i,i} := 1 - \sum_{\substack{j \in \mathsf{S} \\ j \neq i}} \widetilde{P}_{i,j} = P_{i,i} + \sum_{\substack{j \in \mathsf{S}, j \neq i \\ \pi_j < \pi_i}} P_{i,j}\left(1 - \frac{\pi_j}{\pi_i}\right), \qquad i \in \mathsf{S}.
$$

Interpretation

Starting from a state $\bigcirc{i}$, a proposal $\bigcirc{j}$ is generated with probability $P_{i,j}$. This proposal is then accepted if $\pi_j \geqslant \pi_i$, otherwise if $\pi_j < \pi_i$, the proposal is accepted with probability π_j / π_i, and one remains at state $\bigcirc{i}$ with probability $1 - \pi_j / \pi_i$, which can be summarized as follows:

$$
\left\{
\begin{array}{l}
\pi_j \geqslant \pi_i \Rightarrow \text{ accept the proposal } \bigcirc{j}, \\[3mm]
\pi_j < \pi_i \Rightarrow \text{ accept the proposal } \bigcirc{j} \text{ with probability } \pi_j / \pi_i. \text{ Otherwise, keep } \bigcirc{i}.
\end{array}
\right.
$$

Generating Posterior Samples Using MCMC

We consider the prior distribution $\mu = (\mu_i)_{i \in S}$ of a model parameter in the state space S. Given $\mathcal{O}$ a set of observations sampled according to a distribution $(\nu_k)_{k \in \mathcal{O}}$, we are given a likelihood function $l(k|i)$ which represents the probability of observing $k \in \mathcal{O}$ when the system parameter is ⓘ $\in S$, with

$$\nu_k = \sum_{i \in S} l(k|i)\mu_i, \qquad k \in \mathcal{O}. \tag{6.11}$$

The posterior probability distribution $\pi(i|k)$ of being in the state ⓘ given that we observed $k \in \mathcal{O}$ is obtained by the Bayes formula as

$$\pi(i|k) = l(k|i)\frac{\mu_i}{\nu_k}, \qquad i \in S, \ k \in \mathcal{O}. \tag{6.12}$$

Computing the posterior distribution $\pi(i|k)$ and generating the corresponding random samples may require estimating the distribution ν_k, $k \in \mathcal{O}$.

The Markov Chain Monte Carlo method provides an efficient way to generate random samples according to the posterior distribution $\pi(i|k)$. For this, we replace the ratio π_j/π_i in (6.10) with the ratio

$$\frac{\pi(j|k)}{\pi(i|k)} = \frac{\pi(j|k)\nu_k}{\pi(i|k)\nu_k} = \frac{l(k|j)\mu_j}{l(k|i)\mu_i}, \qquad i, j \in S, \ k \in \mathcal{O}, \tag{6.13}$$

which uses the information given by the observation ⓚ. We note that this approach does not rely on the values of $\pi(j|k)$ and $\pi(i|k)$, whose computation through (6.12) would require estimating ν_k via (6.11).

Relation (6.13) shows that the proposal ⓙ generated with probability $P_{i,j}$ is accepted if $\pi(j|k) \geqslant \pi(i|k)$, i.e. if its posterior probability $\pi(j|k)$ given the observation k is higher than the posterior probability $\pi(i|k)$ of the initial state ⓘ. Otherwise, if $\pi(j|k) < \pi(i|k)$ the proposal ⓙ is accepted only with the probability given by (6.13).

Improved versions of the MCMC algorithms include the Hamiltonian Monte Carlo method and the No U-Turn Sampler (NUTS).

Implementation Example

We consider an example on the continuous parameter state space $S := [0, 1]$. Let $N \geqslant 1$, and consider

- a set $\mathcal{O} = \{0, 1\}^N$ of observation values,
- a prior distribution with uniform density $(\mu_\zeta)_{\zeta \in S}$ on the parameter space S,

- a Bernoulli product likelihood distribution with parameter $\zeta \in S$ on $\mathcal{O}$, i.e.

$$l(e_1, \ldots, e_N | \zeta) = \zeta^{e_1 + \cdots + e_N} (1 - \zeta)^{N - (e_1 + \cdots + e_N)},$$

$(e_1, \ldots, e_N) \in \mathcal{O}$.

In this special case, the density $\pi(\zeta | k)$ of the posterior distribution on $S = [0, 1]$ can be explicitly computed for $k = (e_1, \ldots, e_N) \in \mathcal{O}$ as

$$\pi(\zeta | e_1, \ldots, e_N) = l(e_1, \ldots, e_N | \zeta) \frac{\mu_\zeta}{v_{e_1, \ldots, e_N}}$$

$$= \frac{1}{v_{e_1, \ldots, e_N}} \zeta^{e_1 + \cdots + e_N} (1 - \zeta)^{N - (e_1 + \cdots + e_N)}, \quad \zeta \in [0, 1],$$

with the normalization

$$v_{e_1, \ldots, e_N} = \int_0^1 l(e_1, \ldots, e_N | \zeta) d\zeta$$

$$= \int_0^1 \zeta^{e_1 + \cdots + e_N} (1 - \zeta)^{N - (e_1 + \cdots + e_N)} d\zeta$$

$$= B(e_1 + \cdots + e_N + 1, N - (e_1 + \cdots + e_N) + 1),$$

$(e_1, \ldots, e_N) \in \{0, 1\}^N$, where

$$B(e_1 + \cdots + e_N + 1, N - (e_1 + \cdots + e_N) + 1)$$

$$= \frac{(e_1 + \cdots + e_N)! (N - (e_1 + \cdots + e_N))!}{(N + 1)!}$$

is the beta function. The ® code_14_page_163.R and code_15_page_163.R implement the Markov Chain Monte Carlo algorithm using the ® package Stan.

```
1   install.packages("rstan"); install.packages("devtools") # Install RTools as well
    library(devtools); library(lattice); library(rstan)
3   stanmodelcode <- "data {int<lower=0> N;int y[N];}
    parameters {real<lower=0,upper=1> theta;}
5   model {theta ~ uniform(0,1);y ~ bernoulli(theta);}"
    N <- 3;y <- rbinom(N, 1, .3)
7   y <- c(0,0,0,0,1,0,0);N=length(y)
    dat <- list(N = N, y = y); sapply(dat, class)
9   fit <- stan(model_code=stanmodelcode, model_name="Bernoulli-uniform", data=dat,
        iter=2000, chains=1, sample_file='norm.csv', verbose=TRUE) # try iter = 100
    traceplot(fit,inc_warmup = TRUE,col="purple");
11  e <- extract(fit)
    mean(e$theta)
13  densityplot(e$theta, xlim = c(0,1),lwd=2)
```

Although the MCMC algorithms is designed to handle large data sets, for illustration purposes we consider a toy model with $N = 3$ and $y = (0, 1, 0)$. In this case we

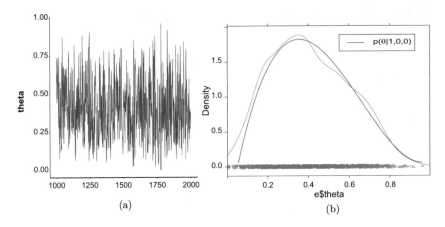

Fig. 6.5 RStan MCMC output. (**a**) Markov chain path. (**b**) Density plot

find $v_{1,0,0} = 2/4! = 1/12$ and the posterior distribution

$$\pi(\zeta|0, 1, 0) = \frac{1}{v_{0,1,0}} l(0, 1, 0|\zeta) = 12\,\zeta(1 - \zeta)^2,$$

as illustrated in Fig. 6.5 using the following Ⓡ code.

```
x=seq(0,1,0.01)
f<-function(x){return (x^(sum(y))*(1-x)^(N-sum(y))/beta(sum(y)+1,N-sum(y)+1))}
par(mar = c(4.3, 2, 2, 3))
plot(x,f(x), lwd=2,col="red")
densityplot(e$theta, xlim=c(0,1),lwd=2)
lines(x,f(x),lwd=2, xlim=c(0,1),col="red")
```

6.3   Transition Bounds and Contractivity

Let P be the transition matrix of a discrete-time Markov chain $(X_n)_{n\geqslant0}$ on $S = \{1, 2, \dots, N\}$.

Definition 6.10 Given two probability distributions $\mu = [\mu_1, \mu_2, \dots, \mu_N]$ and $v = [v_1, \mu_2, \dots, v_N]$ on $\{1, 2, \dots, N\}$, the ℓ^1 distance between μ and v is defined as

$$\|\mu - v\|_1 := \sum_{k=1}^{N} |\mu_k - v_k|.$$

In what follows, for any $A \subset S$ we let

$$\mu(A) = \sum_{k \in A} \mu_k.$$

Definition 6.11 The *total variation distance* between two probability distributions μ and ν on $\{1, 2, \ldots, N\}$ is defined as

$$\|\mu - \nu\|_{\mathrm{TV}} := \underset{A \subset \{1,2,\ldots,N\}}{\mathrm{Max}} |\mu(A) - \nu(A)|. \tag{6.14}$$

In Lemma 6.12 we determine the set A^* on which the maximum of (6.14) is attained.

Lemma 6.12 *Given μ, ν two probability distributions on $\{1, 2, \ldots, N\}$, we have*

$$\|\mu - \nu\|_{\mathrm{TV}} = \mu(A^*) - \nu(A^*) = \sum_{k \in A^*} (\mu_k - \nu_k),$$

where the set $A^* \subset \{1, 2, \ldots, N\}$ is given by

$$A^* := \{k \in \{1, 2, \ldots, N\} \ : \ \nu_k \leqslant \mu_k\}.$$

Proof As the maximum in (6.14) is over a finite number of values, it is attained by A^* provided that

$$\|\mu - \nu\|_{\mathrm{TV}} := \underset{A \subset \{1,2,\ldots,N\}}{\mathrm{Max}} |\mu(A) - \nu(A)| \leqslant |\mu(A^*) - \nu(A^*)|.$$

By construction of the set A^*, we check that for all $A \subset \{1, 2, \ldots, N\}$ we have

$$\mu(A) - \nu(A) = \sum_{k \in A} (\mu_k - \nu_k)$$

$$\leqslant \sum_{k \in A^*} (\mu_k - \nu_k)$$

$$= \mu(A^*) - \nu(A^*),$$

and similarly

$$\mu(A) - \nu(A) = (1 - \mu(A^c)) - (1 - \nu(A^c))$$

$$= -\mu(A^c) + \nu(A^c)$$

$$= -\sum_{k \in A^c} (\mu_k - \nu_k)$$

$$\geqslant - \sum_{k \in A^*} (\mu_k - \nu_k)$$

$$= -(\mu(A^*) - \nu(A^*)),$$

which allows us to conclude. $\square$

The total variation distance is connected to the ℓ^1 distance by the following proposition.

Proposition 6.13 *For any two probability distributions μ and ν on $\{1, 2, \ldots, N\}$, we have*

$$\|\mu - \nu\|_{\mathrm{TV}} = \frac{1}{2}\|\mu - \nu\|_1 = \frac{1}{2} \sum_{k=1}^{N} |\mu_k - \nu_k|.$$

Proof Letting

$$A^* := \{k \in \{1, 2, \ldots, N\} \ : \ \nu_k \leqslant \mu_k\},$$

we have

$$\|\mu - \nu\|_1 = \sum_{k=1}^{N} |\mu_k - \nu_k|$$

$$= \sum_{k \in A^*} |\mu_k - \nu_k| + \sum_{k \in (A^*)^c} |\mu_k - \nu_k|$$

$$= \sum_{k \in A^*} (\mu_k - \nu_k) + \sum_{k \in (A^*)^c} (\nu_k - \mu_k)$$

$$= \sum_{k \in A^*} (\mu_k - \nu_k) + \sum_{k \in (A^*)^c} \nu_k - \sum_{k \in (A^*)^c} \mu_k$$

$$= \sum_{k \in A^*} (\mu_k - \nu_k) + 1 - \sum_{k \in A^*} \nu_k - \left(1 - \sum_{k \in A^*} \mu_k\right)$$

$$= \sum_{k \in A^*} (\mu_k - \nu_k) + \sum_{k \in A^*} (\mu_k - \nu_k)$$

$$= 2 \sum_{k \in A^*} (\mu_k - \nu_k)$$

$$= 2(\mu(A^*) - \nu(A^*))$$

$$= 2\|\mu - \nu\|_{\mathrm{TV}},$$

where the last equality comes from Lemma 6.12. $\square$

The next result is a direct consequence of Proposition 6.13.

Proposition 6.14 *For any two probability distributions μ and v on $\{1, 2, \ldots, N\}$, we always have $\|\mu - v\|_{\mathrm{TV}} \leqslant 1$.*

Proof We have

$$\|\mu - v\|_{\mathrm{TV}} = \frac{1}{2} \sum_{k=1}^{N} |\mu_k - v_k|$$

$$\leqslant \frac{1}{2} \sum_{k=1}^{N} (\mu_k + v_k)$$

$$= \frac{1}{2} \sum_{k=1}^{N} \mu_k + \frac{1}{2} \sum_{k=1}^{N} v_k$$

$$= 1.$$

$\square$

Recall that the vector $\mu P^n = ([\mu P^n]_i)_{i=1,2,\ldots,N}$ denotes the probability distribution of the chain at time $n \in \mathbb{N}$, given it was started with the initial distribution $\mu = [\mu_1, \mu_2, \ldots, \mu_N]$, i.e. we have, using matrix product notation,

$$\mathbb{P}(X_n = i) = \sum_{j=1}^{N} \mathbb{P}(X_n = i \mid X_0 = j)\mathbb{P}(X_0 = j) = \sum_{j=1}^{N} \mu_j [P^n]_{j,i} = [\mu P^n]_i,$$

$i = 1, 2, \ldots, N$. The next lemma presents a contractivity property for the transition matrix P.

Lemma 6.15 *For any two probability distributions $\mu = [\mu_1, \mu_2, \ldots, \mu_N]$ and $v = [v_1, v_2, \ldots, v_N]$ on $\{1, 2, \ldots, N\}$ and any Markov transition matrix P we have*

$$\|\mu P - v P\|_{\mathrm{TV}} \leqslant \|\mu - v\|_{\mathrm{TV}}.$$

Proof Using the triangle inequality

$$\left| \sum_{k=1}^{N} x_k \right| \leqslant \sum_{k=1}^{N} |x_k|, \qquad x_1, x_2, \ldots, x_N \in \mathbb{R},$$

we have

$$
\|\mu P - v P\|_{\mathrm{TV}} = \frac{1}{2} \sum_{j=1}^{N} \left| [\mu P]_j - [v P]_j \right|
$$

$$
= \frac{1}{2} \sum_{j=1}^{N} \left| \sum_{i=1}^{n} \mu_i P_{i,j} - \sum_{i=1}^{n} v_i P_{i,j} \right|
$$

$$
= \frac{1}{2} \sum_{j=1}^{N} \left| \sum_{i=1}^{n} (\mu_i - v_i) P_{i,j} \right|
$$

$$
\leqslant \frac{1}{2} \sum_{j=1}^{N} \sum_{i=1}^{n} \left| (\mu_i - v_i) P_{i,j} \right|
$$

$$
= \frac{1}{2} \sum_{j=1}^{N} \sum_{i=1}^{n} P_{i,j} |\mu_i - v_i|
$$

$$
= \frac{1}{2} \sum_{i=1}^{n} |\mu_i - v_i| \sum_{j=1}^{N} P_{i,j}
$$

$$
= \frac{1}{2} \sum_{i=1}^{n} |\mu_i - v_i|
$$

$$
= \|\mu - v\|_{\mathrm{TV}}.
$$

$\square$

By induction on $n \geqslant 1$, Lemma 6.15 also shows that

$$
\|\mu P^{n+1} - v P^{n+1}\|_{\mathrm{TV}} \leqslant \|\mu P^n - v P^n\|_{\mathrm{TV}} \leqslant \|\mu - v\|_{\mathrm{TV}}, \qquad n \geqslant 1.
$$

When the chain with transition matrix P admits a stationary distribution we obtain the following corollary.

Corollary 6.16 *Assume that the chain $(X_n)_{n \geqslant 0}$ admits a stationary distribution $\pi = [\pi_1, \pi_2, \ldots, \pi_N]$. Then, for any probability distribution $\mu = [\mu_1, \mu_2, \ldots, \mu_N]$ we have*

$$
\|\mu P^{n+1} - \pi\|_{\mathrm{TV}} \leqslant \|\mu P^n - \pi\|_{\mathrm{TV}}, \qquad n \geqslant 0.
$$

Proof Replacing μ and v with μP^n and π in Lemma 6.15, we have

$$
\|\mu P^{n+1} - \pi\|_{\mathrm{TV}} = \|(\mu P^n) P - \pi P\|_{\mathrm{TV}} \leqslant \|\mu P^n - \pi\|_{\mathrm{TV}}, \qquad n \geqslant 0.
$$

$\square$

6.4 Distance to Stationarity

Next, we let

$$d(n) := \underset{\mu \in \mathcal{P}_N}{\text{Max}} \| \mu P^n - \pi \|_{\text{TV}}, \qquad n \geqslant 0,$$

denote the *distance to stationarity* of X_n to $\pi = [\pi_1, \pi_2, \ldots, \pi_N]$.

Lemma 6.17 *The distance to stationarity $d(n)$ is a nonincreasing function, i.e. we have $d(n+1) \leqslant d(n)$, $n \geqslant 0$.*

Proof Letting $\mu \in \mathcal{P}_N$ by Corollary 6.16 we find

$$\| \mu P^{n+1} - \pi P \|_{\text{TV}} \leqslant \| \mu P^n - \pi \|_{\text{TV}}.$$

Taking the maximum over $\mu \in \mathcal{P}_N$ in the above inequality yields

$$
\begin{aligned}
d(n+1) &= \underset{\mu \in \mathcal{P}_N}{\text{Max}} \| \mu P^{n+1} - \pi \|_{\text{TV}} \\
&\leqslant \underset{\mu \in \mathcal{P}_N}{\text{Max}} \| \mu P^n - \pi \|_{\text{TV}} \\
&= d(n), \qquad n \geqslant 0.
\end{aligned}
$$

$\square$

Remark 6.18

(i) If all entries in P are *strictly positive* then the chain is aperiodic and irreducible, and it admits a limiting and stationary distribution. Indeed, the chain is irreducible because all states can communicate in one time step since $P_{i,j} > 0$, $1 \leqslant i, j \leqslant N$. In addition, the chain is aperiodic as all states have period one, given that $P_{i,i} > 0$, $i = 1, 2, \ldots, N$.

(ii) Since the state space is finite, Corollary 1.21 shows that all states are positive recurrent, hence by Corollary 6.7 the chain admits a limiting and a stationary distribution that are equal.

In what follows, we make the following assumption.

Assumption (C) *Assume that the transition matrix P admits an invariant (or stationary) distribution $\pi = [\pi_1, \pi_2, \ldots, \pi_N]$ such that $\pi P = \pi$, and that for some $0 < \theta < 1$ we have*

$$P_{i,j} \geqslant \theta \pi_j, \quad \text{for all } i, j = 1, 2, \ldots, N. \tag{6.15}$$

We also let

$$
\Pi := \begin{bmatrix} \pi \\ \\ \pi \\ \\ \pi \\ \\ \vdots \\ \\ \pi \end{bmatrix} = \begin{bmatrix} \pi_1 & \pi_2 & \pi_3 & \pi_4 & \cdots & \pi_N \\ \pi_1 & \pi_2 & \pi_3 & \pi_4 & \cdots & \pi_N \\ \pi_1 & \pi_2 & \pi_3 & \pi_4 & \cdots & \pi_N \\ \vdots & \vdots & \vdots & \vdots & \ddots & \vdots \\ \pi_1 & \pi_2 & \pi_3 & \pi_4 & \cdots & \pi_N \end{bmatrix},
$$

hence (6.15) reads $P \geqslant \theta \Pi$ using componentwise ordering, and the optimal value of θ may be found as

$$
\theta^* = \min_{1 \leqslant i, j \leqslant N} \frac{P_{i,j}}{\pi_j}.
$$

In addition, since π is a stationary distribution for P we have the relation

$$
\Pi = \Pi P. \tag{6.16}
$$

Lemma 6.19 *Under Assumption (C), for all $0 < \theta < 1$ the matrix*

$$
Q_\theta := \frac{1}{1 - \theta} (P - \theta \Pi)
$$

is the transition matrix of a Markov chain on $S = \{1, 2, \ldots, N\}$ which admits π as stationary distribution. We also note the relation $Q\Pi = \Pi$ for any Markov transition matrix Q.

Proof We note that the matrix Q_θ has nonnegative entries due to Assumption (C), and it can be written as

$$
Q_\theta = \left[[Q_\theta]_{i,j} \right]_{1 \leqslant i, j \leqslant N}
$$

$$
= \begin{bmatrix} [Q_\theta]_{1,1} & [Q_\theta]_{1,2} & \cdots & [Q_\theta]_{1,N} \\ [Q_\theta]_{2,1} & [Q_\theta]_{2,2} & \cdots & [Q_\theta]_{2,N} \\ \vdots & \vdots & \ddots & \vdots \\ [Q_\theta]_{N,1} & [Q_\theta]_{N,2} & \cdots & [Q_\theta]_{N,N} \end{bmatrix}
$$

$$
= \begin{bmatrix}
\frac{1}{1-\theta}(P_{1,1} - \theta\pi_1) & \frac{1}{1-\theta}(P_{1,2} - \theta\pi_2) & \cdots & \frac{1}{1-\theta}(P_{1,N} - \theta\pi_N) \\[2mm]
\frac{1}{1-\theta}(P_{2,1} - \theta\pi_1) & \frac{1}{1-\theta}(P_{2,2} - \theta\pi_2) & \cdots & \frac{1}{1-\theta}(P_{2,N} - \theta\pi_N) \\[2mm]
\vdots & \vdots & \ddots & \vdots \\[2mm]
\frac{1}{1-\theta}(P_{N,1} - \theta\pi_1) & \frac{1}{1-\theta}(P_{N,2} - \theta\pi_2) & \cdots & \frac{1}{1-\theta}(P_{N,N} - \theta\pi_N)
\end{bmatrix}.
$$

Clearly, all entries of Q_θ are nonnegative due to the condition

$$
P_{i,j} \geqslant \theta\pi_j, \qquad i, j = 1, 2, \ldots, N.
$$

In addition, for all $i = 1, 2, \ldots, N$ we have

$$
\sum_{j=1}^{N} [Q_\theta]_{i,j} = \frac{1}{1-\theta} \sum_{j=1}^{N} (P_{i,j} - \theta\Pi_{i,j})
$$

$$
= \frac{1}{1-\theta} \sum_{j=1}^{N} (P_{i,j} - \theta\pi_j)
$$

$$
= \frac{1}{1-\theta} \sum_{j=1}^{N} P_{i,j} - \frac{\theta}{1-\theta} \sum_{j=1}^{N} \pi_j
$$

$$
= \frac{1}{1-\theta} - \frac{\theta}{1-\theta}
$$

$$
= 1, \qquad 0 < \theta < 1,
$$

and we conclude that Q_θ is a Markov transition matrix. The stationarity of π with respect to Q_θ follows from

$$
\pi Q_\theta = \frac{1}{1-\theta}(\pi P - \theta\pi\Pi) = \frac{\pi - \theta\pi}{1-\theta} = \pi.
$$

$\square$

Lemma 6.20 *We have the relation*

$$
P^n - \Pi = (1 - \theta)^n (Q_\theta^n - \Pi), \qquad n \geqslant 0. \tag{6.17}
$$

Proof This statement is proved by induction on $n \in \mathbb{N}$. Clearly, the property holds for $n = 0$, and for $n = 1$ by the definition of Q_θ. Next, assume that

$$
P^n = \Pi + (1 - \theta)^n (Q_\theta^n - \Pi)
$$

for some $n \geqslant 1$. Noting that the condition $\pi P = \pi$ implies $\Pi P = \Pi$ and using the relation $P = \Pi + (1 - \theta)(Q_\theta - \Pi)$, we have

$$
\begin{aligned}
P^{n+1} &= P^n P \\
&= \big(\Pi + (1 - \theta)^n (Q_\theta^n - \Pi)\big) P \\
&= \Pi P + (1 - \theta)^n Q_\theta^n P - (1 - \theta)^n \Pi P \\
&= \Pi + (1 - \theta)^n Q_\theta^n P - (1 - \theta)^n \Pi \\
&= \Pi + (1 - \theta)^n Q_\theta^n \big(\Pi + (1 - \theta)(Q_\theta - \Pi)\big) - (1 - \theta)^n \Pi \\
&= \Pi + \theta(1 - \theta)^n Q_\theta^n \Pi + (1 - \theta)^{n+1} Q_\theta^{n+1} - (1 - \theta)^n \Pi.
\end{aligned}
$$

Next, we note that we have $R\Pi = \Pi$ for any Markov transition matrix R, hence $P\Pi = \Pi^2 = \Pi$, and

$$
Q_\theta \Pi = \frac{1}{1 - \theta}(P - \theta \Pi)\Pi = \frac{1}{1 - \theta}(P\Pi - \theta \Pi^2) = \frac{\Pi - \theta \Pi}{1 - \theta} = \Pi,
$$

hence $Q_\theta \Pi = \Pi$, and more generally $Q_\theta^n \Pi = \Pi$, $n \geqslant 1$. Therefore, we have

$$
\begin{aligned}
P^{n+1} &= \Pi + \theta(1 - \theta)^n Q_\theta^n \Pi + (1 - \theta)^{n+1} Q_\theta^{n+1} - (1 - \theta)^n \Pi \\
&= \Pi + \theta(1 - \theta)^n \Pi + (1 - \theta)^{n+1} Q_\theta^{n+1} - (1 - \theta)^n \Pi \\
&= \Pi + (1 - \theta)^{n+1} Q_\theta^{n+1} - (1 - \theta)^{n+1} \Pi \\
&= \Pi + (1 - \theta)^{n+1} \big(Q_\theta^{n+1} - \Pi\big),
\end{aligned}
$$

which allows us to conclude by induction. $\square$

We refer to Theorem 4.9 in Levin et al. (2009) for the next result.

Proposition 6.21 *Under Assumption (C), given any initial distribution μ the total variation distance between the distribution μP^n of the chain at time n and its stationary distribution $\pi = [\pi_1, \pi_2, \ldots, \pi_N]$ satisfies*

$$
\big\|\mu P^n - \pi\big\|_{\mathrm{TV}} \leqslant (1 - \theta)^n, \qquad n \geqslant 1, \quad \mu \in \mathcal{P}_N.
$$

As a consequence, we have

$$
d(n) \leqslant (1 - \theta)^n, \qquad n \geqslant 1.
$$

Proof Let $\mu \in \mathcal{P}_N$. Relation (6.17) shows that

$$
\big\|\mu P^n - \pi\big\|_{\mathrm{TV}} = \big\|\mu P^n - \mu \Pi\big\|_{\mathrm{TV}}
$$

$$= \frac{1}{2} \sum_{j=1}^{N} \left| [\mu(P^n - \Pi)]_j \right|$$

$$= \frac{1}{2} \sum_{j=1}^{N} (1-\theta)^n \left| [\mu Q_\theta^n - \pi]_j \right|$$

$$= \frac{(1-\theta)^n}{2} \sum_{j=1}^{N} \left| [\mu Q_\theta^n]_j - \pi_j \right|$$

$$= (1-\theta)^n \left\| \mu Q_\theta^n - \pi \right\|_{\text{TV}}$$

$$\leqslant (1-\theta)^n, \qquad n \geqslant 0,$$

where we applied Proposition 6.14, since $\Pi_{k,\cdot} = \pi$ is a probability distribution and the same holds for $[Q_\theta^n]_{k,\cdot}$ for all $k = 1, 2, \ldots, N$ by Lemma 6.19. Finally, we find

$$d(n) = \operatorname*{Max}_{\mu \in \mathcal{P}_N} \left\| \mu P^n - \pi \right\|_{\text{TV}} \leqslant (1-\theta)^n, \qquad n \geqslant 0.$$

$\square$

The relation

$$\left\| \mu P^n - \pi \right\|_{\text{TV}} = (1-\theta)^n \left\| \mu Q_\theta^n - \pi \right\|_{\text{TV}}, \qquad n \geqslant 0,$$

also shows that, in total variation distance, at each time step the chain associated to P converges faster (by a factor $1 - \theta$) to π than the chain associated to Q_θ.

Remark 6.22 Proposition 6.21 shows that any stationary distribution satisfying the condition $P_{i,j} \geqslant \theta \pi_j$, $i, j = 1, 2, \ldots, N$, admits the limiting distribution

$$\pi_j := \lim_{n \to \infty} \mathbb{P}(X_n = j \mid X_0 = i) = \lim_{n \to \infty} [P^n]_{i,j}, \qquad i, j \in \mathsf{S},$$

independently of the initial state $i \in \mathsf{S}$.

Remark 6.22 applies in particular when $P_{i,j} > 0$, $i, j = 1, 2, \ldots, N$, in which case the chain is irreducible and aperiodic, and admits a unique limiting and stationary distribution. More generally, the result holds when P is *regular*, i.e. when there exists $n \geqslant 1$ such that $[P^n]_{i,j} > 0$ for all $i, j = 1, 2, \ldots, N$, cf. § 4.3–4.5 of Levin et al. (2009).

Note that if the transition matrix $P = (P_{i,j})_{1 \leqslant i, j \leqslant N}$ has strictly positive entries, it can be shown as in Propositions 4–5 of Bryan and Leise (2006) that for any initial distribution μ we have

$$\left\| \mu P^n - \pi \right\|_1 \leqslant c^n \|\mu - \pi\|_1, \qquad n \geqslant 0,$$

with

$$c := \underset{i=1,2,\dots,N}{\text{Max}} \left| 1 - 2 \underset{j=1,2,\dots,N}{\min} P_{i,j} \right|,$$

see Exercise 6.7

6.5   Mixing Times

The *mixing time* of the chain with transition matrix P is defined as

$$t_{\text{mix}}^{\alpha} := \min\{n \geqslant 0 \; : \; d(n) \leqslant \alpha\},$$

for some threshold $\alpha \in (0, 1)$. In what follows, we let

$$\lceil x \rceil = \min\{n \in \mathbb{Z} \; : \; x \leqslant n\}$$

denote the integer *ceiling* of $x \in \mathbb{R}$.

Proposition 6.23 *The mixing time t_{mix}^{α} of the chain associated to P satisfies the exponential convergence rate*

$$t_{\text{mix}}^{\alpha} \leqslant \left\lceil \frac{\log \alpha}{\log(1 - \theta)} \right\rceil.$$

Proof If $t_{\text{mix}}^{\alpha} = 0$ the inequality is clearly satisfied, so that we can suppose that $t_{\text{mix}}^{\alpha} \geqslant 1$. By Lemma 6.17 the *distance to stationarity* $d(n)$ is a nonincreasing function, hence by the definition of t_{mix}^{α} and Proposition 6.21 we have

$$\alpha < d(t_{\text{mix}}^{\alpha} - 1) \leqslant (1 - \theta)^{t_{\text{mix}}^{\alpha} - 1},$$

hence

$$\log \alpha < \log d(t_{\text{mix}}^{\alpha} - 1) \leqslant \log \left((1 - \theta)^{t_{\text{mix}}^{\alpha} - 1} \right) = (t_{\text{mix}}^{\alpha} - 1) \log(1 - \theta).$$

Dividing the above inequality by $\log(1 - \theta) < 0$ yields

$$t_{\text{mix}}^{\alpha} - 1 \leqslant \frac{\log d(t_{\text{mix}}^{\alpha} - 1)}{\log(1 - \theta)} < \frac{\log \alpha}{\log(1 - \theta)}.$$

Hence, we have

$$t_{\text{mix}}^{\alpha} < 1 + \frac{\log \alpha}{\log(1 - \theta)},$$

which yields

$$t_{\text{mix}}^{\alpha} < 1 + \left\lceil \frac{\log \alpha}{\log(1 - \theta)} \right\rceil,$$

and finally

$$t_{\text{mix}}^{\alpha} \leqslant \left\lceil \frac{\log \alpha}{\log(1 - \theta)} \right\rceil.$$

$\square$

The condition $P_{i,j} \geqslant \theta \pi_j$, $i, j = 1, 2, 3$, reads

$$P = \begin{bmatrix} \dfrac{2}{3} & \dfrac{1}{6} & \dfrac{1}{6} \\[8pt] \dfrac{1}{3} & \dfrac{1}{2} & \dfrac{1}{6} \\[8pt] \dfrac{1}{6} & \dfrac{2}{3} & \dfrac{1}{6} \end{bmatrix} \geqslant \theta \begin{bmatrix} \dfrac{11}{24} & \dfrac{9}{24} & \dfrac{4}{24} \\[8pt] \dfrac{11}{24} & \dfrac{9}{24} & \dfrac{4}{24} \\[8pt] \dfrac{11}{24} & \dfrac{9}{24} & \dfrac{4}{24} \end{bmatrix}$$

or

$$\left[\frac{P_{i,j}}{\pi_j} \right]_{1 \leqslant i,j \leqslant 3} = \begin{bmatrix} \dfrac{48}{33} & \dfrac{12}{27} & 1 \\[8pt] \dfrac{24}{33} & \dfrac{12}{9} & 1 \\[8pt] \dfrac{4}{11} & \dfrac{48}{27} & 1 \end{bmatrix} \geqslant \begin{bmatrix} \theta & \theta & \theta \\ \theta & \theta & \theta \\ \theta & \theta & \theta \end{bmatrix}, \tag{6.18}$$

where the inequality is understood componentwise, hence the optimal (largest possible) value of θ such that $\theta \leqslant P_{i,j}/\pi_j$, $i, j = 1, 2, 3$, is

$$\theta^* = \min_{1 \leqslant i,j \leqslant 3} \frac{P_{i,j}}{\pi_j} = \frac{4}{11}.$$

Taking $\alpha = 1/4$ and $\theta = 4/11$, we have

$$t_{\text{mix}}^{\alpha} \leqslant \left\lceil \frac{\log 1/4}{\log(1 - \theta)} \right\rceil = \left\lceil \frac{\log 1/4}{\log 7/11} \right\rceil = \lceil 3.067 \rceil = 4.$$

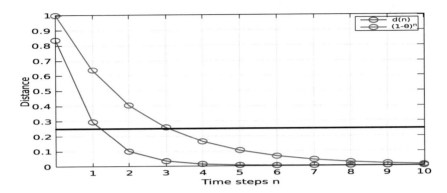

Fig. 6.6 Graphs of distance to stationarity $d(n)$ and upper bound $(1 - \theta)^n$ with $\alpha = 1/4$

We check from Fig. 6.6 that the actual value of the mixing time is $t^{\alpha}_{\text{mix}} = 2$, where we estimate $d(n)$ as

$$d(n) := \underset{k=1,2,\ldots,N}{\text{Max}} \left\| [P^n]_{k,\cdot} - \pi \right\|_{\text{TV}}, \qquad n \geqslant 0.$$

The value of $d(0)$ is the maximum distance between π and all deterministic initial distributions starting from states $k = 1, 2, \ldots, N$.

Below is the Matlab/Octave code used to generate Fig. 6.6.

```
   P = [2/3,1/6,1/6; 1/3,1/2,1/6; 1/6,2/3,1/6;]
2  pi = [11/24,9/24,4/25];theta = 4/11;
   for n = 1:11
4  y(n)=n-1;u(n)=0.25;z(n)=(1-theta)^(n-1);distance(n) = 0;
   for k = 1:3;d = mpower(P,n-1)(k,1:3) - pi;dist=0;
6  for i = 1:3;dist = dist + 0.5*abs(d(i));end
   distance(n) = max(distance(n) ,dist);end;end
8  graphics_toolkit("gnuplot");
   plot(y,distance,'-bo','LineWidth',3,y,z,'-ro','LineWidth',3,y,u,'-k', 'LineWidth',5)
10 legend('d(n)','(1-\theta)^n')
   set (gca, 'xtick', 1:10,"fontsize", 12)
12 set (gca, 'ytick', 0:0.1:1,"fontsize", 12)
   grid on
14 xlabel('Time steps n',"fontsize", 12);ylabel('Distance',"fontsize", 12)
```

Coupling

We close this chapter with a general bound on the distance between the distributions of two arbitrary discrete-time random sequences $(X_n)_{n \geqslant 0}$ and $(Y_n)_{n \geqslant 0}$ on a state space $\mathbb{S}$, for some random time called τ the *coupling time* of $(X_n)_{n \geqslant 0}$ and $(Y_n)_{n \geqslant 0}$, such that

$$X_n = Y_n, \qquad n \geqslant \tau.$$

Proposition 6.24 *For all $n \in \mathbb{N}$, we have*

$$\underset{x \in S}{\mathrm{Sup}} \, |\mathbb{P}(X_n = x) - \mathbb{P}(Y_n = x)| \leqslant \mathbb{P}(\tau > n), \qquad n \geqslant 0.$$

Proof By the law of total probability, for all $x \in S$ and $n \geqslant 0$ we have

$$
\begin{aligned}
\mathbb{P}(X_n = x) &= \mathbb{P}(\{X_n = x\} \cap \{\tau \leqslant n\}) + \mathbb{P}(\{X_n = x\} \cap \{\tau > n\}) \\
&= \mathbb{P}(\{Y_n = x\} \cap \{\tau \leqslant n\}) + \mathbb{P}(\{X_n = x\} \cap \{\tau > n\}) \\
&\leqslant \mathbb{P}(Y_n = x) + \mathbb{P}(\tau > n).
\end{aligned}
$$

Similarly to the above, we have

$$
\begin{aligned}
\mathbb{P}(Y_n = x) &= \mathbb{P}(\{Y_n = x\} \cap \{\tau \leqslant n\}) + \mathbb{P}(\{Y_n = x\} \cap \{\tau > n\}) \\
&= \mathbb{P}(\{X_n = x\} \cap \{\tau \leqslant n\}) + \mathbb{P}(\{Y_n = x\} \cap \{\tau > n\}) \\
&\leqslant \mathbb{P}(X_n = x) + \mathbb{P}(\tau > n),
\end{aligned}
$$

hence

$$-\mathbb{P}(\tau > n) \leqslant \mathbb{P}(X_n = x) - \mathbb{P}(Y_n = x) \leqslant \mathbb{P}(\tau > n), \quad x \in S, \quad n \geqslant 0,$$

which leads to

$$\underset{x \in S}{\mathrm{Sup}} \, |\mathbb{P}(X_n = x) - \mathbb{P}(Y_n = x)| \leqslant \mathbb{P}(\tau > n), \qquad n \geqslant 0.$$

$\square$

See Exercise 6.12-(f) for an application of the coupling technique to random shuffling.

Notes

See e.g. § 4.3–4.5 of Levin et al. (2009) for further reading.

Exercises

Exercise 6.1 Compute the limiting and stationary distributions of the Markov chain $(Y_k)_{k \geqslant 0}$ with transition matrix (3.12).

Exercise 6.2 Find the stationary distribution $[\pi_0, \pi_1]$ of the two-state Markov chain on $S = \{0, 1\}$ with transition probability matrix

$$
P = \begin{array}{c} 0 \\ 1 \end{array} \!\! \begin{array}{cc} 0 & 1 \\ \left[\begin{array}{cc} 1/3 & 2/3 \\ 2/3 & 1/3 \end{array} \right]. \end{array}
$$

Exercise 6.3 Let $(Y_k)_{k\in\mathbb{N}}$ denote the Markov chain considered in § 3.3.

(a) Is the chain $(Y_k)_{k\in\mathbb{N}}$ reducible? Find its communicating classes.
(b) Find the limiting distribution, and the possible stationary distributions of the chain $(Y_k)_{k\in\mathbb{N}}$.

Exercise 6.4 Consider a two-state $\{0, 1\}$-valued Markov chain $(X_n)_{n\geqslant 0}$ on the state space with transition matrix

$$
P = \begin{array}{c} 0 \\ 1 \end{array} \!\! \begin{array}{cc} 0 & 1 \\ \left[\begin{array}{cc} 1 - a & a \\ b & 1 - b \end{array} \right], \end{array}
$$

where $a, b \in (0, 1)$. This question is to be treated via explicit computations for two-state Markov chains, without referring to general results.

(a) Give the stationary distribution $\pi = (\pi_0, \pi_1)$ of the chain $(X_n)_{n\geqslant 0}$.
(b) Compute the mean return times $\mu_0(0)$, $\mu_1(1)$ and the mean hitting times $h_0(1)$, $h_1(0)$ of the chain $(X_n)_{n\geqslant 0}$.
(c) Compute the conditional expected values $\mathbb{E}[\tau \mid X_0 = 0]$ and $\mathbb{E}[\tau \mid X_0 = 1]$ of the cycle length

$$
\tau := \inf\{l > 1 \ : \ X_l = X_1\}.
$$

(d) Compute the four expected values

$$
\mathbb{E}\left[\sum_{l=1}^{\tau - 1} \mathbb{1}_{\{X_l = i\}} \,\middle|\, X_0 = j \right], \qquad i, j = 0, 1.
$$

(e) Show that for any initial distribution $(\mathbb{P}(X_0 = 0), \mathbb{P}(X_0 = 1))$ we have

$$
\pi_0 = \frac{\mathbb{E}\left[\sum_{l=1}^{\tau - 1} \mathbb{1}_{\{X_l = 0\}} \right]}{\mathbb{E}[\tau - 1]}, \qquad \pi_1 = \frac{\mathbb{E}\left[\sum_{l=1}^{\tau - 1} \mathbb{1}_{\{X_l = 1\}} \right]}{\mathbb{E}[\tau - 1]}.
$$

Exercise 6.5 Given $(X_n)_{n\geqslant 0}$ an irreducible Markov chain with transition matrix P and stationary distribution $\pi = [\pi_1, \pi_2, \ldots, \pi_N]$ on the state space $S = \{1, 2, \ldots, N\}$, consider the distances to stationarity defined as

$$d(n) := \underset{\mu \in \mathcal{P}_N}{\text{Max}} \|\mu P^n - \pi\|_1 \quad \text{and} \quad \widehat{d}(n) := \underset{k=1,2,\ldots,N}{\text{Max}} \|[P^n]_{k,\cdot} - \pi\|_1, \quad n \geqslant 0,$$

where $\mathcal{P}_N$ is the set of probability measures on $\{1, \ldots, N\}$ and

$$\|\mu - \nu\|_1 := \sum_{k=1}^{N} |\mu_k - \nu_k|$$

denotes the ℓ^1 distance between any two probability distributions $\mu = [\mu_1, \mu_2, \ldots, \mu_N]$, $\nu = [\nu_1, \mu_2, \ldots, \nu_N]$ on S.

(a) Show that $\widehat{d}(n) \leqslant d(n)$, $n \geqslant 0$.
(b) Show that $d(n) \leqslant \widehat{d}(n)$, $n \geqslant 0$.

Exercise 6.6 (Aldous and Diaconis (1986) and Jonasson (2009)) Let $(X_n)_{n\geqslant 1}$ denote a Markov chain on a finite state space S, and let $\tau \geqslant 0$ denote a random time such that the distribution π of X_n given $\{\tau \leqslant n\}$ does not depend on $n \geqslant 0$, i.e.

$$\mathbb{P}(X_n \in A \mid \tau \leqslant n) = \pi(A), \qquad A \subset S, \quad n \geqslant 0.$$

(a) Show that

$$\mathbb{P}(X_n \in A) = \pi(A) + (\mathbb{P}(X_n \in A \mid \tau > n) - \pi(A))\mathbb{P}(\tau > n),$$

$A \subset S, n \geqslant 0$.
Hint. Split $\mathbb{P}(X_n \in A)$ as

$$\mathbb{P}(X_n \in A) = \mathbb{P}(X_n \in A \text{ and } \tau \leqslant n) + \mathbb{P}(X_n \in A \text{ and } \tau > n).$$

(b) Show the total variation distance bound

$$\|\mathbb{P}(X_n \in \cdot) - \pi(\cdot)\|_{\text{TV}} := \underset{A \subset S}{\text{Sup}} |\mathbb{P}(X_n \in A) - \pi(A)| \leqslant \mathbb{P}(\tau > n),$$

between π and the distribution of X_n, $n \geqslant 0$.
Hint. Use the inequalities

$$-1 \leqslant a - 1 \leqslant a - b \leqslant 1 - b \leqslant 1, \qquad a, b \in [0, 1].$$

(c) Give an example of a random time such that the distribution π of X_n given $\{\tau \leqslant n\}$ does not depend on $n \geqslant 0$.

Exercise 6.7 (Bryan and Leise (2006)) Let $M = (M_{i,j})_{1 \leqslant i,j \leqslant n}$ denote a *column-stochastic* matrix, i.e. M is such that

$$\sum_{i=1}^{n} M_{i,j} = 1, \qquad j = 1, 2, \ldots, n,$$

and assume that M has strictly positive entries, i.e.

$$M_{i,j} > 0, \qquad i, j = 1, 2, \ldots, n.$$

We let $\|x\|_1 = \sum_{k=1}^{n} |x_k|$ denote the ℓ^1 norm of $x = (x_1, \ldots, x_n) \in \mathbb{R}^n$. Prove the following statements using only Markov chain reasoning.

(a) Show that M admits 1 as (right) eigenvalue and that the corresponding eigenspace has dimension 1.
(b) Show that there exists a unique vector $y \in \mathbb{R}^n$ with positive components such that $My = y$ with $\|y\|_1 = 1$, which can be computed as $y = \lim_{k \to \infty} M^k x_0$ for any initial guess x_0 with positive components such that $\|x_0\|_1 = 1$.

Exercise 6.8 Consider an irreducible positive recurrent Markov chain $(X_n)_{n \geqslant 0}$ with unique stationary distribution π on a state space S, and let

$$\tau_x := \inf\{n \geqslant 1 \ : \ X_n = x\}$$

denote the first return time to state $x \in S$.

(a) Let

$$R_n^x := \sum_{k=1}^{n} \mathbb{1}_{\{X_k = x\}}$$

denote the number of returns to state $x \in S$ from time 1 to time n. Show that the stationary distribution $\pi = (\pi_x)_{x \in S}$ satisfies

$$\pi_x = \lim_{n \to \infty} \frac{\mathbb{E}[R_n^x]}{n}, \qquad x \in S.$$

Hint. Show that the limit satisfies $\pi = \pi P$.

(b) Let

$$N_{x,y} := \sum_{n=1}^{\tau_x} \mathbf{1}_{\{X_n = y\}}$$

denote the number of visits to state y before the first return to state x. Show that we have

$$\pi_y = \frac{\mathbb{E}[N_{x,y} \mid X_0 = x]}{\mathbb{E}[\tau_x \mid X_0 = x]}, \qquad x, y \in S.$$

Hint. Use the law of large numbers for regenerative processes.

(c) Show that $N_{x,y}$ has a geometric distribution, and find its parameter in terms of $\alpha_{x,y} := \mathbb{P}(N_{x,y} \geqslant 1 \mid X_0 = x)$ and $\alpha_{y,x} := \mathbb{P}(N_{y,x} \geqslant 1 \mid X_0 = y)$, $x, y \in S$.

(d) Find a relation between π_x, π_y, $\alpha_{x,y}$, $\alpha_{y,x}$.

Hint. Recall that we have

$$\pi_x = \frac{1}{\mathbb{E}[\tau_x \mid X_0 = x]}, \qquad x \in S,$$

and

$$\sum_{k \geqslant 1} k r^{k-1} = \frac{1}{(1-r)^2},$$

for any $r \in [0, 1)$, see (B.4).

Problem 6.9 Consider a two-state Markov chain $(X_n)_{n \geqslant 0}$ on $S = \{0, 1\}$, with transition matrix

$$P = \begin{array}{c} \\ 0 \\ 1 \end{array} \!\! \begin{array}{c} \quad 0 \qquad 1 \\ \left[\begin{array}{cc} 1-a & a \\ b & 1-b \end{array} \right], \end{array}$$

where $a, b \in (0, 1)$.

(a) Find the lowest eigenvalue λ of P.

(b) Find the stationary distribution (π_0, π_1) of the chain $(X_n)_{n \geqslant 0}$.

(c) Show by induction on $n \geqslant 0$ that

$$\begin{bmatrix} \mathbb{E}\left[\exp\left(t \sum_{k=1}^{n} X_k \right) \mid X_0 = 0 \right] \\ \mathbb{E}\left[\exp\left(t \sum_{k=1}^{n} X_k \right) \mid X_0 = 1 \right] \end{bmatrix} = \left(\begin{bmatrix} 1-a & ae^t \\ b & (1-b)e^t \end{bmatrix} \right)^n \begin{bmatrix} 1 \\ 1 \end{bmatrix}.$$

$n \geqslant 0$, $t \in \mathbb{R}$. In the sequel, we assume that $(X_n)_{n \geqslant 0}$ is started in its *stationary distribution*, i.e.

$$\mathbb{P}(X_0 = 0) = \pi_0, \qquad \mathbb{P}(X_0 = 1) = \pi_1.$$

(d) Show that for all $n \geqslant 1$ we have

$$\mathbb{E}\left[\exp\left(t \sum_{k=1}^{n} X_k \right) \right]$$

$$= [\sqrt{\pi_0}, \sqrt{\pi_1} e^{t/2}] \left(\begin{bmatrix} \lambda + (1 - \lambda)\pi_0 & (1 - \lambda)e^{t/2}\sqrt{\pi_0 \pi_1} \\ (1 - \lambda)e^{t/2}\sqrt{\pi_0 \pi_1} & (\lambda + (1 - \lambda)\pi_1)e^t \end{bmatrix} \right)^{n-1}$$

$$\times \begin{bmatrix} \sqrt{\pi_0} \\ \sqrt{\pi_1} e^{t/2} \end{bmatrix}.$$

Hint. Diagonalize P as

$$\begin{bmatrix} 1 - a & a \\ b & (1 - b) \end{bmatrix} = \begin{bmatrix} \frac{1}{\sqrt{\pi_0}} & 0 \\ 0 & \frac{1}{\sqrt{\pi_1}} \end{bmatrix} \begin{bmatrix} \sqrt{\pi_0} & -\sqrt{\pi_1} \\ \sqrt{\pi_1} & \sqrt{\pi_0} \end{bmatrix} \begin{bmatrix} 1 & 0 \\ 0 & \lambda \end{bmatrix}$$

$$\times \begin{bmatrix} \sqrt{\pi_0} & \sqrt{\pi_1} \\ -\sqrt{\pi_1} & \sqrt{\pi_0} \end{bmatrix} \begin{bmatrix} \sqrt{\pi_0} & 0 \\ 0 & \sqrt{\pi_1} \end{bmatrix},$$

and use the fact that

$$\begin{bmatrix} 1 - a & ae^t \\ b & (1 - b)e^t \end{bmatrix} = \begin{bmatrix} 1 - a & a \\ b & 1 - b \end{bmatrix} \begin{bmatrix} 1 & 0 \\ 0 & e^t \end{bmatrix}.$$

(e) Find the largest eigenvalue $\mu(t)$ of the matrix

$$M(t) := \begin{bmatrix} \lambda + (1 - \lambda)\pi_0 & (1 - \lambda)e^{t/2}\sqrt{\pi_0 \pi_1} \\ (1 - \lambda)e^{t/2}\sqrt{\pi_0 \pi_1} & (\lambda + (1 - \lambda)\pi_1)e^t \end{bmatrix}.$$

In the sequel, we assume that $\lambda \geqslant 0$.

(f) Show that for all $n \geqslant 0$ and $t \in \mathbb{R}_+$ we have

$$\mathbb{E}\left[\exp\left(t \sum_{k=1}^{n} X_k\right)\right] \leqslant (\pi_0 + \pi_1 e^t)(\mu(t))^{n-1} \leqslant (\mu(t))^n.$$

Hint. Use e.g. Proposition 9 in Foucart (2010).

(g) Using the Markov inequality, show that

$$\mathbb{P}\left(\frac{1}{n} \sum_{k=1}^{n} (X_k - \pi_1) \geqslant z\right) \leqslant e^{-n((\pi_1 + z)t - \log \mu(t))}, \qquad z > 0, \quad t > 0.$$

(h) Show that for all $n \geqslant 1$ we have

$$\mathbb{P}\left(\frac{1}{n} \sum_{k=1}^{n} (X_k - \pi_1) \geqslant z\right) \leqslant \exp\left(-2\frac{1-\lambda}{1+\lambda} n z^2\right), \qquad z > 0.$$

Hint. Find the value $t(x)$ of $t > 0$ that maximizes $t \mapsto xt - \log \mu(t)$ for x fixed in $(0, 1)$, and then show that

$$\frac{xt(x) - \log \mu(t(x))}{(x - \pi_1)^2} \geqslant 2\frac{1-\lambda}{1+\lambda}, \qquad x \in (0, 1).$$

Problem 6.10 Let $(X_n)_{n \geqslant 0}$ denote an irreducible aperiodic Markov chain on a finite state space S, with transition matrix $P = (P_{i,j})_{i,j \in S}$ and stationary distribution $\pi = (\pi_i)_{i \in S}$. We let

$$R_n^i := \sum_{k=1}^{n} \mathbb{1}_{\{X_k = i\}}$$

denote the number of returns to state $i \in S$ from time 1 to time n. Recall that by Exercise 6.8-(a) the stationary distribution $\pi = (\pi_i)_{i \in S}$ satisfies

$$\pi_i = \lim_{n \to \infty} \frac{\mathbb{E}[R_n^i]}{n}, \qquad i \in S. \tag{6.19}$$

(a) Define the sequence $(\tau_k)_{k \geqslant 1}$ recursively as

$$\tau_1 := \inf\{l > 1 \ : \ X_l = X_1\},$$

and

$$\tau_k := \inf\{l > \tau_{k-1} \ : \ X_l = X_1\}, \qquad k \geqslant 2.$$

Show, using e.g. Theorem 31 p. 15 of Freedman (1983) and the law of large numbers for regenerative processes, see Corollary 14 p. 106 of Serfozo (2009), that

$$\pi_i = \frac{\mathbb{E}\left[\sum_{j=1}^{\tau_1 - 1} \mathbb{1}_{\{X_j = i\}}\right]}{\mathbb{E}[\tau_1 - 1]}, \qquad i \in S.$$

(b) Let τ be a stopping time for $\mathcal{F}_n := \sigma(X_0, \ldots, X_n)$, $n \geqslant 0$, with $\mathbb{E}[\tau] < \infty$. By writing

$$T := \inf\{l > \tau \ : \ X_l = X_1\}$$

as $T = \tau_\kappa$ where κ is a stopping time[1] for $(\mathcal{F}_{\tau_k})_{k \geqslant 1}$, show that

$$\pi_i = \frac{\mathbb{E}\left[\sum_{j=1}^{T-1} \mathbb{1}_{\{X_j = i\}}\right]}{\mathbb{E}[T - 1]}, \qquad i \in S.$$

Hint. Use e.g. Theorem 2 of Chewi (2017).

Problem 6.11 (Problem 4.2 Continued) We consider an N-arm bandit in which arm $n°i$ is modeled by a two-state Markov chain $(X_n^{(i)})_{n \geqslant 0}$ on $S := \{0, 1\}$, with transition matrix $P^{(i)}$ and stationary distribution $(\pi_0^{(i)}, \pi_1^{(i)})$, $i = 1, \ldots, N$, ordered as $\pi_1^{(1)} \leqslant \cdots \leqslant \pi_1^{(N)}$. Given an $\{1, \ldots, N\}$-valued policy $(\alpha_k)_{k \geqslant 1}$, we let

$$T_n^{(i,\alpha)} := \sum_{k=1}^{n} \mathbb{1}_{\{\alpha_k = i\}}, \qquad i = 1, 2, \ldots, N,$$

denote the number of times the arm i is selected by the policy $(\alpha_k)_{k \geqslant 1}$ until time $n \geqslant 1$. The reward of arm $n°i$ after it has been pulled $n \geqslant 1$ times is $X_n^{(i)}$, and the *regret* $\mathcal{R}_n^\alpha$ at time n of the policy $(\alpha_k)_{k \geqslant 1}$ is given by

$$\mathcal{R}_n^\alpha := n\pi_1^{(N)} - \mathbb{E}\left[\sum_{i=1}^{N} \sum_{k=1}^{T_n^{(i,\alpha)}} X_k^{(i)}\right], \qquad n \geqslant 1.$$

---

[1] See e.g. § 2 of Chewi (2017).

(a) Bounded regret (Problem 6.9 continued).

(i) Show that for any stopping time τ for $\mathcal{F}_n := \sigma(X_0^{(i)}, \ldots, X_n^{(i)})$, $n \geqslant 0$, letting $R_\tau^{(i)} := \sum_{k=1}^\tau \mathbb{1}_{\{X_k^{(i)}=1\}}$ denote the number of returns to state $\textcircled{1}$ until time τ by the chain $(X_n^{(i)})_{n \geqslant 1}$, we have

$$\left| \mathbb{E}[R_\tau^{(i)}] - \pi_1^{(i)} \mathbb{E}[\tau] \right| \leqslant \mathbb{E}[T - \tau], \quad i = 1, \ldots, N,$$

where $T := \inf\{l > \tau : X_l = X_1\}$.
Hint. Use the relations

$$R_{T-1}^{(i)} - (T - \tau) \leqslant R_T^{(i)} - (T - \tau) \leqslant R_\tau^{(i)} \leqslant R_{T-1}^{(i)}$$

in the notation of Question (b) of Problem 6.10.
(ii) Show that

$$\left| \mathbb{E}\left[\sum_{i=1}^N \sum_{k=1}^{T_n^{(i,\alpha)}} X_k^{(i)} - \sum_{i=1}^N \pi_1^{(i)} T_n^{(i,\alpha)} \right] \right| \leqslant 2 \sum_{i=1}^N \underset{l,j \in S}{\text{Max}} \, \mu_l^{(i)}(j), \quad n > N,$$

where $\mu_l^{(i)}(j)$ denotes the first return time of state $j \in S$ from state $l \in S$ by the chain $(X_n^{(i)})_{n \geqslant 0}$.
(iii) Show that the *regret* $\mathcal{R}_n^\alpha$ of the policy $(\alpha_k)_{k \geqslant 1}$ is bounded as

$$\mathcal{R}_n^\alpha \leqslant \overline{\mathcal{R}}_n^\alpha + K, \quad n > N,$$

for some constant $K > 0$ independent of $n \geqslant 1$, where $\overline{\mathcal{R}}_n^\alpha$ is the *modified* regret defined as

$$\overline{\mathcal{R}}_n^\alpha := n \pi_1^{(N)} - \mathbb{E}\left[\sum_{i=1}^N \pi_1^{(i)} T_n^{(i,\alpha)} \right], \quad n \geqslant 1.$$

(b) Learning at the $\log n$ speed. Let

$$\widehat{m}_n^{(i,\alpha)} := \frac{1}{T_n^{(i,\alpha)}} \sum_{k=1}^{T_n^{(i,\alpha)}} X_k^{(i)}$$

denote the sample average reward obtained from arm $n°i$ until time $n \geqslant 1$ under the policy $(\alpha_k)_{k \geqslant 1}$.

Given $L > 0$, we define the policy $(\alpha_n^*)_{n \geqslant 1}$ by $\alpha_n^* := n$ for $n = 1, \ldots, N$, and for $n > N$ we let α_n^* be the index $i \in \{1, \ldots, N\}$ that maximizes the quantity

$$\widehat{m}_{n-1}^{(i,\alpha^*)} + \sqrt{\frac{L \log n}{T_{n-1}^{(i,\alpha^*)}}}.$$

(i) Let $1 \leqslant i < N$ and $n \geqslant N$. Show by contradiction that if $\alpha_n^* = i$, then at least one of the following three conditions must hold:

$$\begin{cases} \widehat{m}_{n-1}^{(N,\alpha^*)} + \sqrt{\dfrac{L \log n}{T_{n-1}^{(N,\alpha^*)}}} \leqslant \pi_1^{(N)}, \\[4mm] \widehat{m}_{n-1}^{(i,\alpha^*)} > \pi_1^{(i)} + \sqrt{\dfrac{L \log n}{T_{n-1}^{(i,\alpha^*)}}}, \\[4mm] T_{n-1}^{(i,\alpha^*)} < \dfrac{4L \log n}{(\pi_1^{(N)} - \pi_1^{(i)})^2}. \end{cases}$$

(ii) Show that letting $\widehat{n}_i := 4L(\log n)/(\pi_1^{(N)} - \pi_1^{(i)})^2$, we have

$$\mathbb{E}[T_n^{(i,\alpha^*)}] \leqslant \widehat{n}_i + \sum_{\widehat{n}_i < k \leqslant n} \left(\mathbb{P}\left(\widehat{m}_{k-1}^{(N,\alpha^*)} + \sqrt{\frac{L \log k}{T_{k-1}^{(N,\alpha^*)}}} \leqslant \pi_1^{(N)} \right) \right.$$

$$\left. + \mathbb{P}\left(\widehat{m}_{k-1}^{(i,\alpha^*)} > \pi_1^{(i)} + \sqrt{\frac{L \log k}{T_{k-1}^{(i,\alpha^*)}}} \right) \right),$$

$1 \leqslant i < N, n \geqslant N.$

(iii) Letting λ_i denote the smallest eigenvalue of $P^{(i)}$, we assume that $\min_{1 \leqslant i \leqslant N} \lambda_i \geqslant 0$, let $\lambda := \mathrm{Max}_{1 \leqslant i \leqslant N} \lambda_i$, and assume that $L > (1 + \lambda)/(1 - \lambda)$.
Show that

$$\mathbb{P}\left(\widehat{m}_{k-1}^{(N,\alpha^*)} + \sqrt{\frac{L \log k}{T_{k-1}^{(N,\alpha^*)}}} \leqslant \pi_1^{(N)} \right) \leqslant \frac{1}{k^{2L(1-\lambda)/(1+\lambda)-1}}$$

and

$$\mathbb{P}\left(\widehat{m}_{k-1}^{(i,\alpha^*)} > \pi_1^{(i)} + \sqrt{\frac{L \log k}{T_{k-1}^{(i,\alpha^*)}}} \right) \leqslant \frac{1}{k^{2L(1-\lambda)/(1+\lambda)-1}},$$

$i = 1, \ldots, N, k > N.$

Fig. 6.7 Top to random shuffling

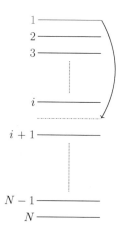

Hint. Apply the result of Question (A)-(8) of Assignment 1.

(iv) Show that the modified regret can be bounded for any $L > (1+\lambda)/(1-\lambda)$ by

$$\overline{\mathcal{R}}_n^{\alpha*} \leqslant \sum_{i=1}^{N-1} \frac{\pi_1^{(N)} - \pi_1^{(i)}}{L(1-\lambda)/(1+\lambda) - 1} + (\log n) \sum_{i=1}^{N-1} \frac{4L}{\pi_1^{(N)} - \pi_1^{(i)}}, \qquad n > N.$$

Hint. Use a comparison argument between series and integrals.

Problem 6.12 (Aldous and Diaconis (1986) and Jonasson (2009)) *Random shuffling* is applied to a deck of $N = 52$ cards by inserting the top card back into the deck at a random location $i \in \{1, \ldots, N\}$ chosen uniformly among $N = 52$ possible positions (Fig. 6.7).

More formally, consider the Markov chain $(X_n)_{n \geqslant 0}$ on the group

$$S_N = \{(e_1, \ldots, e_N) : e_1, \ldots, e_N \in \{1, \ldots, N\}, \; e_i \neq e_j, \; 1 \leqslant i \neq j \leqslant N\}$$

of $N!$ permutations of $(1, \ldots, N)$, built by applying the cycle permutation of indexes

$$(1, 2, \ldots, i) \mapsto (2, \ldots, i, 1)$$

to X_n for some uniformly chosen $i \in \{1, \ldots, N\}$ if $i \geqslant 2$, or the identity if $i = 1$. The transition matrix P of the chain is given by

$$\mathbb{P}(X_{n+1} = (e_1, \ldots, e_N) \mid X_n = (e_1, \ldots, e_N)) := \frac{1}{N},$$

and

$$\mathbb{P}(X_{n+1} = (e_2, \ldots e_i, e_1, e_{i+1}, \ldots, e_N)) \mid X_n = (e_1, \ldots, e_N)) := \frac{1}{N},$$

$i = 2, \ldots, N$, with $P_{\sigma,\eta} := 0$ in all other cases, with $\sigma, \eta \in S_N$.

At time 0 we choose to start with the initial condition $X_0 := (1, \ldots, N)$. We also let $T_0 := 0$, and for $k = 1, \ldots, N - 1$ we denote by T_k the first time the original bottom card has moved up to the rank $N - k$ in the deck. Note that at time T_{N-1}, the original bottom card should have moved to the top of the deck.

(a) Find the probability distribution

$$\mathbb{P}(T_l - T_{l-1} = m), \quad m \geqslant 1, \quad \text{for} \quad l = 1, \ldots, N - 1.$$

  Hint. This is a geometric distribution. Find its parameter depending on $l = 1, \ldots, N - 1$.

(b) Find the mean time $\mathbb{E}[T_k]$ it takes until the original bottom card has moved to the position $N - k, k = 1, \ldots, N - 1$.
  Hint. Use the telescoping identity

$$T_k = (T_k - T_{k-1}) + (T_{k-1} - T_{k-2}) + \cdots + (T_2 - T_1) + (T_1 - T_0).$$

(c) Compute $\mathrm{Var}[T_{N-1}]$, and show that $\mathrm{Var}[T_{N-1}] \leqslant CN^2$ for some constant $C > 0$.
  Hint. The random variables $T_k - T_{k-1}, k = 1, \ldots, N - 1$, are independent.

(d) Show that for any $a > 0$ we have

$$\lim_{N \to \infty} \mathbb{P}(T_{N-1} > (1 + a)N \log N) = 0.$$

  Hint. Use Chebyshev's inequality

$$\mathbb{P}(Z - \mathbb{E}[Z] \geqslant x) \leqslant \frac{1}{x^2} \mathrm{Var}[Z], \quad x > 0,$$

  and the bound

$$\sum_{k=1}^{N-1} \frac{1}{k} \leqslant 1 + \log N, \quad N \geqslant 1.$$

(e) What is the distribution of X_n given that $n > T_{N-1}$?
  Hint. The answer is intuitive. No proof is required.

(f) Based on the answers to Questions (d)-(e) and the coupling argument of Proposition 6.24, find the convergence rate of the distribution of $(X_n)_{n \geqslant 0}$ to the uniform distribution.

Problem 6.13 (Levin et al. (2009)) Convergence to equilibrium. In this problem we derive quantitative bounds for the convergence of a Markov chain to its stationary distribution π. Let P be the transition matrix of a discrete-time Markov chain $(X_n)_{n \geqslant 0}$ on $\mathbb{S} = \{1, 2, \ldots, N\}$. Given two probability distributions $\mu = [\mu_1, \mu_2, \ldots, \mu_N]$ and $\nu = [\nu_1, \nu_2, \ldots, \nu_N]$ on $\{1, 2, \ldots, N\}$, the *total variation distance* between μ and ν is defined as

$$\|\mu - \nu\|_{\mathrm{TV}} := \frac{1}{2} \sum_{k=1}^{N} |\mu_k - \nu_k|.$$

Recall that the vector $\mu P^n = ([\mu P^n]_i)_{i=1,2,\ldots,N}$ denotes the probability distribution of the chain at time $n \in \mathbb{N}$, given it was started with the initial distribution $\mu = [\mu_1, \mu_2, \ldots, \mu_N]$, i.e. we have, using matrix product notation,

$$\mathbb{P}(X_n = i) = \sum_{j=1}^{N} \mathbb{P}(X_n = i \mid X_0 = j)\mathbb{P}(X_0 = j) = \sum_{j=1}^{N} \mu_j [P^n]_{j,i} = [\mu P^n]_i,$$

$i = 1, 2, \ldots, N$.

(a) Show that for any two probability distributions $\mu = [\mu_1, \mu_2, \ldots, \mu_N]$ and $\nu = [\nu_1, \nu_2, \ldots, \nu_N]$ on $\{1, 2, \ldots, N\}$ we always have $\|\mu - \nu\|_{\mathrm{TV}} \leqslant 1$.

(b) Show that for any two probability distributions $\mu = [\mu_1, \mu_2, \ldots, \mu_N]$ and $\nu = [\nu_1, \nu_2, \ldots, \nu_N]$ on $\{1, 2, \ldots, N\}$ and any Markov transition matrix P we have

$$\|\mu P - \nu P\|_{\mathrm{TV}} \leqslant \|\mu - \nu\|_{\mathrm{TV}}.$$

Hint: Use the triangle inequality

$$\left| \sum_{k=1}^{n} x_k \right| \leqslant \sum_{k=1}^{n} |x_k|, \qquad x_1, x_2, \ldots, x_n \in \mathbb{R}.$$

(c) Assume that the chain with transition matrix P admits a stationary distribution $\pi = [\pi_1, \pi_2, \ldots, \pi_N]$. Show that for any probability distribution $\mu = [\mu_1, \mu_2, \ldots, \mu_N]$ we have

$$\|\mu P^{n+1} - \pi\|_{\mathrm{TV}} \leqslant \|\mu P^n - \pi\|_{\mathrm{TV}}, \qquad n \geqslant 0.$$

(d) Show that the *distance to stationarity*, defined as

$$d(n) := \operatorname*{Max}_{k=1,2,\ldots,N} \|[P^n]_{k,\cdot} - \pi\|_{\mathrm{TV}}, \qquad n \geqslant 0,$$

satisfies $d(n+1) \leqslant d(n), n \in \mathbb{N}$.

(e) Assume that all entries of P are *strictly positive*. Explain why the chain is aperiodic and irreducible, and why it admits a limiting and stationary distribution.

In what follows we assume that P admits an invariant (or stationary) distribution $\pi = [\pi_1, \pi_2, \ldots, \pi_N]$ such that $\pi P = \pi$, and that

$$P_{i,j} \geqslant \theta \pi_j, \quad \text{for all } i, j = 1, 2, \ldots, N, \tag{6.20}$$

for some $0 < \theta < 1$. We also let

$$\Pi := \begin{bmatrix} \pi_1 & \pi_2 & \pi_3 & \pi_4 & \cdots & \pi_N \\ \pi_1 & \pi_2 & \pi_3 & \pi_4 & \cdots & \pi_N \\ \pi_1 & \pi_2 & \pi_3 & \pi_4 & \cdots & \pi_N \\ \vdots & \vdots & \vdots & \vdots & \ddots & \vdots \\ \pi_1 & \pi_2 & \pi_3 & \pi_4 & \cdots & \pi_N \end{bmatrix},$$

hence (6.20) reads $P \geqslant \theta \Pi$.

(f) Show that for all $0 < \theta < 1$ the matrix

$$Q_\theta := \frac{1}{1 - \theta}(P - \theta \Pi)$$

is the transition matrix of a Markov chain on $S = \{1, 2, \ldots, N\}$.

(g) Show by induction on $n \in \mathbb{N}$ that we have

$$P^n - \Pi = (1 - \theta)^n (Q_\theta^n - \Pi), \quad n \in \mathbb{N}.$$

(h) Show that given any $X_0 = k = 1, 2, \ldots, N$ the total variation distance between the distribution

$$[P^n]_{k,\cdot} = ([P^n]_{k,1}, \ldots, [P^n]_{k,N})$$
$$= [\mathbb{P}(X_n = 1 \mid X_0 = k), \ldots, \mathbb{P}(X_n = N \mid X_0 = k)]$$

of the chain at time n and the stationary distribution $\pi = [\pi_1, \pi_2, \ldots, \pi_N]$ satisfies

$$\left\| [P^n]_{k,\cdot} - \pi \right\|_{\mathrm{TV}} \leqslant (1 - \theta)^n, \quad n \geqslant 1, \quad k = 1, 2, \ldots, N.$$

Conclude that we have $d(n) \leqslant (1 - \theta)^n, n \geqslant 1$.

(i) Show that the *mixing time* of the chain with transition matrix P, defined as

$$t_{\text{mix}} := \min\{n \geqslant 0 \,:\, d(n) \leqslant 1/4\},$$

satisfies

$$t_{\text{mix}} \leqslant \left\lceil \frac{\log 1/4}{\log(1 - \theta)} \right\rceil.$$

(j) Find the optimal value of θ satisfying the condition $P_{i,j} \geqslant \theta \pi_j$ for all $i, j = 1, 2, \ldots, N$ for the chain of Exercise 4.12 in Privault (2018), with $N = 3$.

Problem 6.14 (Lezaud (1998)) Consider an *irreducible, reversible,*[2] Markov chain $(X_n)_{n \geqslant 0}$ with transition matrix $P = (P_{i,j})_{1 \leqslant i,j \leqslant d}$ and admitting a stationary distribution π on the finite state space $S = \{1, 2, \ldots, d\}$. For any function $f : \mathbb{R} \to \mathbb{R}$, we let D_f denote the diagonal matrix

$$D_f = \begin{bmatrix} f(1) & 0 & 0 & 0 \cdots & 0 \\ 0 & f(2) & 0 & 0 \cdots & 0 \\ 0 & 0 & f(3) & 0 \cdots & 0 \\ \vdots & \vdots & \vdots & \vdots \ddots & \vdots \\ 0 & 0 & 0 & 0 \cdots & f(d) \end{bmatrix}.$$

We use the scalar product $\langle \cdot, \cdot \rangle$ and norm $\| \cdot \|$ on $\mathbb{R}^d$ defined as

$$\langle u, v \rangle := \sum_{l=1}^{d} u(l) v(l) \pi_l, \quad \|u\|^2 := \sum_{l=1}^{d} |u(l)|^2 \pi_l, \quad u, v \in \mathbb{R}^d,$$

with the Cauchy-Schwarz inequality

$$|\langle u, v \rangle| \leqslant \|u\| \cdot \|v\|, \quad u, v \in \mathbb{R}^d.$$

Recall that the norm $\| \cdot \|$ also defines a matrix norm on $\mathbb{R}^{d \times d}$ as

$$\|M\| = \underset{\substack{u \in \mathbb{R}^d \\ u \neq 0}}{\text{Sup}} \frac{\|Mu\|}{\|u\|} = \underset{\|u\|=1}{\text{Sup}} \|Mu\|, \quad M \in \mathbb{R}^{d \times d}.$$

In what follows, we assume that $(X_n)_{n \geqslant 0}$ is started with π as initial distribution, and $f : \{1, \ldots, d\} \to \mathbb{R}$ denotes any function such that $\|f\|_\infty \leqslant 1$ and $\mathbb{E}[f(X_n)] = 0$, $n \geqslant 0$.

---

[2] i.e. $\pi_i P_{i,j} = \pi_j P_{j,i}$, $i, j = 1, \ldots, d$.

(a) Show that 1 is an eigenvalue of single multiplicity for P, and give its eigenvector.

   Hint. Use the irreducibility of $(X_n)_{n \geqslant 0}$ and the Perron-Frobenius theorem.

(b) Write down the matrix Π of the orthogonal[3] projection operator on the eigenvector of P with eigenvalue 1.

(c) Show by induction on $n \geqslant 0$ that for any state $k \in \{1, \ldots, d\}$ and $\alpha \in \mathbb{R}$, we have

$$\mathbb{E}\left[\exp\left(\alpha \sum_{l=1}^{n} f(X_l)\right) \,\middle|\, X_0 = k\right] = \sum_{l=1}^{d} [(Pe^{\alpha D_f})^n]_{k,l}, \qquad n \geqslant 0.$$

Remark This extends Question (b) of Problem 6.9.

(d) Show that for any $\alpha \geqslant 0$ and $\gamma \geqslant 0$ we have

$$\mathbb{P}\left(\sum_{l=1}^{n} f(X_l) \geqslant n\gamma \,\middle|\, X_0 = k\right) \leqslant e^{-\alpha\gamma n} \sum_{l=1}^{d} [(Pe^{\alpha D_f})^n]_{k,l}, \qquad n \geqslant 0.$$

   Hint. Use the Chernoff argument.

(e) Letting $\lambda_0(\alpha)$ denote the largest eigenvalue of $Pe^{\alpha D_f}$, show that for all $\alpha \geqslant 0$ we have

$$\sum_{k,l=1}^{d} \pi_k [(Pe^{\alpha D_f})^n]_{k,l} \leqslant e^\alpha (\lambda_0(\alpha))^n, \qquad n \geqslant 0. \qquad (6.21)$$

   Hints. (*i*) Write the left hand side of (6.21) as a scalar product and use the Cauchy-Schwarz inequality. (*ii*) Note that

$$Pe^{\alpha D_f} = e^{-\alpha D_f/2} e^{\alpha D_f/2} P e^{\alpha D_f/2} e^{\alpha D_f/2}$$

   is similar to a self-adjoint operator. (*iii*) Apply e.g. Proposition 9 in Foucart (2010).

(f) Show that for any $\alpha \geqslant 0$ and $\gamma \geqslant 0$ we have

$$\mathbb{P}\left(\frac{1}{n}\sum_{l=1}^{n} f(X_l) \geqslant \gamma\right) \leqslant e^{\alpha - n(\alpha\gamma - \log \lambda_0(\alpha))}, \qquad n \geqslant 0.$$

---

[3] Orthogonality is with respect to the scalar product $\langle \cdot, \cdot \rangle$.

(g) Show that for any matrix M we have the relation

$$\text{tr}(\Pi P D_f^n M D_f^m) = \text{tr}(\Pi D_f^n M D_f^m) = \langle f^n, M f^m \rangle, \qquad n, m \geqslant 0. \qquad (6.22)$$

(h) Show that $\lambda_0(\alpha)$ can be expanded as the power series

$$\lambda_0(\alpha) = 1 + \sum_{n \geqslant 1} c_n \alpha^n$$

in the parameter α, with $c_1 = 0$ and

$$c_n = \sum_{p=1}^{n} \frac{(-1)^{p-1}}{p}$$

$$\sum_{\substack{v_1 + \cdots + v_p = n \\ k_1 + \cdots + k_p = p-1 \\ v_1 \geqslant 1, \ldots, v_p \geqslant 1 \\ k_1 \geqslant 0, \ldots, k_p \geqslant 0}} \frac{1}{v_1! \cdots v_p!} \langle f^{v_1}, S^{k_1'} P(D_f)^{v_2} \cdots S^{k_{p-2}'} P(D_f)^{v_{p-1}} S^{k_{p-1}'} P f^{v_p} \rangle,$$

$n \geqslant 2$.

Hints. (i) Apply Relations II-(2.1) and II-(2.31) in Kato (1995) to the expansion

$$P e^{\alpha D_f} = \sum_{n \geqslant 0} \alpha^n P \frac{(D_f)^n}{n!}$$

using the reduced resolvent $S := (P - I)^{-1}(I - \Pi)$, see II-(2.10)–(2.12) and p. 74 line -1. (ii) Use the fact that at least one of $k_1, \ldots, k_p$ must be zero in II-(2.31) of Kato (1995), and denote the non-zero indexes by $k_1', \ldots, k_{p-1}'$. (iii) Use (6.22). (iv) Use $\text{tr}(AB) = \text{tr}(BA)$.

(i) Compute $\displaystyle\sum_{\substack{k_1 + \cdots + k_p = p-1 \\ k_1 \geqslant 0, \ldots, k_p \geqslant 0}} 1$. *Hint.* We have $\displaystyle\sum_{\substack{v_1 + \cdots + v_p = n \\ v_1 \geqslant 1, \ldots, v_p \geqslant 1}} 1 = \binom{n-1}{p-1}$.

(j) Show that $c_n \leqslant (5/(1 - \lambda_1))^{n-1}/5$, $n \geqslant 2$, where λ_1 is the second largest eigenvalue of P.

Hints. (i) Use the inequalities $n! \geqslant 2^{n-1}$ and $4^n \geqslant \binom{2n}{n}\sqrt{\pi n}$, $n \geqslant 1$, Proposition 9 in Foucart (2010), and the Cauchy-Schwarz inequality. (ii) Show that $\|I - \Pi\| \leqslant 1$. (iii) Note that $P - I$ is invertible on $\text{Im}(I - \Pi)$. (iv) Show that

$$\sum_{p=0}^{n-1} \binom{n-1}{p} \frac{x^p}{p+1} = \frac{(1+x)^n - 1}{nx} \leqslant \frac{(1+x)^n}{nx}.$$

(k) Show that for all $\gamma \geqslant 0$ and $n \geqslant 0$ we have

$$\mathbb{P}\left(\frac{1}{n}\sum_{l=1}^{n}f(X_l) \geqslant \gamma\right) \leqslant \exp\left(\frac{1-\lambda_1}{5} - n\gamma\alpha + \frac{n\alpha^2}{1-\lambda_1-5\alpha}\right),$$

$\alpha \in [0, (1-\lambda_1)/5)$.

Hint. Use the inequality $\log(1+x) \leqslant x, x > 0$.

(l) Show that for all $\gamma \geqslant 0$ and $n \geqslant 0$ we have

$$\mathbb{P}\left(\frac{1}{n}\sum_{l=1}^{n}f(X_l) \geqslant \gamma\right) \leqslant e^{(1-\lambda_1)/5}\exp\left(-(1-\lambda_1)\frac{n\gamma^2}{12}\right).$$

Hint. Minimize the upper bound of Question (k) over $\alpha \in [0, (1-\lambda_1)/5)$.

Chapter 7
The Ising Model

This chapter presents the Ising model and studies its long run behavior via its limiting and stationary distribution. Applications of the Ising model can be found in spatial statistics, image analysis and segmentation, opinion studies, urban segregation, language change, metal alloys, magnetic materials, liquid/gas coexistence, phase transitions, plasmas, cell membranes in biophysics, etc.

7.1 Construction

The one-dimensional Ising model is built on the state space $S := \{-1, +1\}^N$ made of elements $z = (z_k)_{1 \leqslant k \leqslant N} \in S$ whose components $z_k \in \{-1, 1\}$, $k = 1, 2, \ldots N$, are called *spins*. The state space S has cardinality 2^N. For example, $2^{100} = 1.26 \times 10^{30}$, see Fig. 7.1.

In the sequel, we write $z = \pm$ to mean that z can take the values $+1$ of -1, i.e. $z \in \{-1, +1\}$. We consider a Markov chain $(Z_n)_{n \geqslant 0}$ *on the state space* $S = \{-1, +1\}^N$, whose transitions from an initial configuration $Z_0 = z = (z_k)_{1 \leqslant k \leqslant N}$ to a new configuration $Z_1 = \tilde{z} = (\tilde{z}_k)_{1 \leqslant k \leqslant N} \in S$ are defined as follows. Let $p \in (0, 1)$ and $q := 1 - p$.

First, randomly pick a component z_k in $z = (z_k)_{1 \leqslant k \leqslant N}$ with probability $1/N$, $k = 1, 2, \ldots, N$, and then consider the following cases:

(i) if $(z_{k-1}, z_{k+1}) = (-1, +1)$ or $(z_{k-1}, z_{k+1}) = (+1, -1)$:
$\Rightarrow$ flip the sign of z_k, i.e. set $\tilde{z}_k := \pm z_k$ with probability $1/2$,
(ii) if $(z_{k-1}, z_{k+1}) = (+1, +1)$:
$\Rightarrow$ set $\tilde{z}_k := +1$ with probability $p > 0$, and $\tilde{z}_k := -1$ with probability $q > 0$.
(iii) if $(z_{k-1}, z_{k+1}) = (-1, -1)$:
$\Rightarrow$ set $\tilde{z}_k := -1$ with probability $p > 0$, and $\tilde{z}_k := +1$ with probability $q > 0$,

© The Author(s), under exclusive license to Springer Nature Switzerland AG 2024
N. Privault, *Discrete Stochastic Processes*, Springer Undergraduate
Mathematics Series, https://doi.org/10.1007/978-3-031-65820-4_7

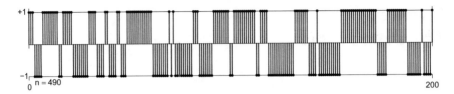

Fig. 7.1 Simulation of the Ising model with $N = 199$, $p = 0.98$, and $z_0 = z_{N+1} = +1$

Fig. 7.2 Simulation of the Ising model with $N = 199$, $p - 0.02$, and $z_0 = z_{N+1} = +1$

where $p + q = 1$. The probabilities p and q can be respectively viewed as the probabilities of "agreeing", resp. "disagreeing" with two neighbors who share the same opinion. The boundary conditions z_0 and z_{N+1} can be arbitrarily specified as in Fig. 7.2, which can be plotted using the ℝ code_16_page_196.R.

The next proposition allows us to formulate the transition probabilities of the chain $(Z_n)_{n \geqslant 0}$ in closed form. For any $z = (z_1, \ldots, z_N) \in \{-1, +1\}^N$ and $k = 1, 2, \ldots, N$, we let

$$\bar{z}^k := (z_1, \ldots, z_{k-1}, -z_k, z_{k+1}, \ldots, z_N) \tag{7.1}$$

denotes the transformation of the state $z \in \mathbb{S}$ obtained after flipping its k-th component z_k, $k = 1, 2, \ldots, N$.

Proposition 7.1 *The transition probabilities*

$$\mathbb{P}(Z_1 = (z_1, \ldots, z_{k-1}, -z_k, z_{k+1}, \ldots, z_N) \mid Z_0 = z)$$

given that $Z_0 = z = (z_1, \ldots, z_N)$ take the general form

$$\mathbb{P}(Z_1 = \bar{z}^k \mid Z_0 = z) = \frac{1}{N(1 + (p/q)^{z_k(z_{k-1} + z_{k+1})/2})}, \quad k = 1, 2, \ldots, N. \tag{7.2}$$

Proof The formula (7.2) follows from computing the transition probabilities

$$\mathbb{P}(Z_1 = (z_1, \ldots, z_{k-1}, -z_k, z_{k+1}, \ldots, z_N) \mid Z_0 = z), \quad k = 1, 2, \ldots, N,$$

given that $Z_0 = z = (z_k)_{1 \leqslant k \leqslant N}$, in the following cases:

(i) $(z_{k-1}, z_k, z_{k+1}) = (-1, \pm 1, +1)$ or $(z_{k-1}, z_k, z_{k+1}) = (+1, \pm 1, -1)$,
(ii) $(z_{k-1}, z_k, z_{k+1}) = (+1, +1, +1)$ or $(z_{k-1}, z_k, z_{k+1}) = (-1, -1, -1)$,
(iii) $(z_{k-1}, z_k, z_{k+1}) = (+, -1, +1)$ or $(z_{k-1}, z_k, z_{k+1}) = (-1, +1, -1)$,

$k = 1, 2, \ldots, N$. In order to conclude, we note that $z_k(z_{k-1} + z_{k+1})/2$ can only take the three possible values $-1, 0, +1$, and treat all cases separately.   □

From (7.2) we can also confirm the relation

$$\mathbb{P}(Z_1 = \bar{z}^k \mid Z_0 = z) + \mathbb{P}(Z_1 = z \mid Z_0 = \bar{z}^k)$$

$$= \frac{1}{N(1 + (p/q)^{z_k(z_{k-1}+z_{k+1})/2})} + \frac{1}{N(1 + (p/q)^{-z_k(z_{k-1}+z_{k+1})/2})}$$

$$= \frac{1}{N}, \quad k = 1, 2, \ldots, N.$$

Example

Taking $N = 3$ and setting $z_0 = z_4 = -1$, i.e. $(z_0, z_1, z_2, z_3, z_4)$ takes the form

$$(z_0, z_1, z_2, z_3, z_4) = (-, \pm, \pm, \pm, -),$$

we find that the transition probability matrix P of $(Z_n)_{n \geqslant 0}$ on the state space $S = \{---, --+, -+-, --+, +--, -+, ++-, +++\}$ is given by

	$---$	$--+$	$-+-$	$-++$	$+--$	$+-+$	$++-$	$+++$
$---$	p	$q/3$	$q/3$	0	$q/3$	0	0	0
$--+$	$p/3$	$1/2$	0	$1/6$	0	$q/3$	0	0
$-+-$	$p/3$	0	$(1+q)/3$	$1/6$	0	0	$1/6$	0
$-++$	0	$1/6$	$1/6$	$1/2$	0	0	0	$1/6$
$+--$	$p/3$	0	0	0	$1/2$	$q/3$	$1/6$	0
$+-+$	0	$p/3$	0	0	$p/3$	q	0	$p/3$
$++-$	0	0	$1/6$	0	$1/6$	0	$1/2$	$1/6$
$+++$	0	0	0	$1/6$	0	$q/3$	$1/6$	$(1+p)/3$

$P =$

For example, since at most one spin may be flipped at any time step and given that $z_0 = z_4 = -1$, we check that

$$\mathbb{P}(Z_1 = ---\mid Z_0 = ---) = \frac{1}{3} \times p + \frac{1}{3} \times p + \frac{1}{3} \times p = p,$$

$$\mathbb{P}(Z_1 = --+\mid Z_0 = ---) = \frac{1}{3} \times 0 + \frac{1}{3} \times 0 + \frac{1}{3} \times q = \frac{q}{3},$$

$$\mathbb{P}(Z_1 = -++\mid Z_0 = --+) = \frac{1}{3} \times 0 + \frac{1}{3} \times \frac{1}{2} + \frac{1}{3} \times 0 = \frac{1}{6},$$

$$\mathbb{P}(Z_1 = +++\mid Z_0 = +++) = \frac{1}{3} \times \frac{1}{2} + \frac{1}{3} \times p + \frac{1}{3} \times \frac{1}{2} = \frac{1+p}{3},$$

etc. When $N = 3$, the chain has the following graph:

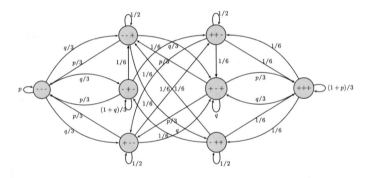

7.2   Irreducibility, Aperiodicity and Recurrence

Aperiodicity

By construction the chain $(Z_n)_{n \geqslant 0}$ is aperiodic since every state has a returning loop because

$$\mathbb{P}(Z_1 = z \mid Z_0 = z) \geqslant \min(p, q) > 0, \qquad z \in \mathbb{S}.$$

More precisely, we can compute $\mathbb{P}(Z_1 = z \mid Z_0 = z)$ for all $z \in S$ using the complement rule, Relation (7.2), and the law of total probability, as

$$\mathbb{P}(Z_1 = z \mid Z_0 = z) = 1 - \sum_{k=1}^{N} \mathbb{P}(Z_1 = \bar{z}^k \mid Z_0 = z)$$

$$= 1 - \frac{1}{N} \sum_{k=1}^{N} \frac{1}{1 + (p/q)^{z_k(z_{k-1} + z_{k+1})/2}}$$

$$= \frac{1}{N} \sum_{k=1}^{N} \left(1 - \frac{1}{1 + (p/q)^{z_k(z_{k-1} + z_{k+1})/2}} \right)$$

$$= \frac{1}{N} \sum_{k=1}^{N} \frac{(p/q)^{z_k(z_{k-1} + z_{k+1})/2}}{1 + (p/q)^{z_k(z_{k-1} + z_{k+1})/2}}$$

$$= \frac{1}{N} \sum_{k=1}^{N} \frac{1}{1 + (q/p)^{z_k(z_{k-1} + z_{k+1})/2}}$$

$$> 0, \qquad z \in S.$$

Irreducibility

The chain is irreducible because starting from any configuration $z = (z_k)_{1 \leqslant k \leqslant N} \in S$ we can reach any other configuration $\hat{z} = (\hat{z}_k)_{1 \leqslant k \leqslant N} \in S$ in a finite number of time steps. In order to check this, we can for example count the number of spins in $z = (z_k)_{1 \leqslant k \leqslant N}$ that differ from the spins in $\hat{z} = (\hat{z}_k)_{1 \leqslant k \leqslant N}$ and flip them one by one until we reach $\hat{z} = (\hat{z}_k)_{1 \leqslant k \leqslant N}$.

Alternatively, we could also enumerate all possible 2^N configurations by flipping one spin at a time, starting from $z = (+1, +1, \ldots, +1)$ until we reach $z = (-1, -1, \ldots, -1)$, and back from $z = (-1, -1, \ldots, -1)$ to $z = (+1, +1, \ldots, +1)$.

Recurrence

The chain has a finite state space of cardinality 2^N and it is irreducible, hence it is positive recurrent by Corollary 1.21. Since in addition the chain is aperiodic, by Theorem 6.6 it admits a limiting distribution and a stationary distribution which coincide.

7.3   Limiting and Stationary Distributions

The chain has a finite state space of cardinality 2^N, it is aperiodic and positive recurrent, hence bye.g. Theorem 6.2 it admits a limiting distribution independent of its initial state, and a unique stationary distribution $(\pi_z)_{z \in S}$ solution of $\pi = \pi P$, which is known to coincide with its limiting distribution. In particular, we have

$$\lim_{n \to \infty} \mathbb{P}(Z_n = z \mid Z_0 = \tilde{z}) = \lim_{n \to \infty} [P^n]_{\tilde{z}, z} = \pi_z, \qquad z, \tilde{z} \in S.$$

In Lemma 7.2, Relation (7.3) is a version of the *detailed balance* condition (6.8), according to Lemma 6.4.

Lemma 7.2 *Any probability distribution $(\pi_z)_{z \in S}$ on S satisfying the relation*

$$\frac{\pi_{\tilde{z}^k}}{\pi_z} = \frac{\mathbb{P}(Z_1 = \tilde{z}^k \mid Z_0 = z)}{\mathbb{P}(Z_1 = z \mid Z_0 = \tilde{z}^k)}, \qquad k = 1, 2, \ldots, N, \quad z \in S, \tag{7.3}$$

where $\tilde{z}^k$ is defined in (7.1), is a stationary distribution *for the chain $(Z_n)_{n \geqslant 0}$, i.e. we have $\pi = \pi P$ and*

$$\big(\mathbb{P}(Z_0 = z) = \pi_z, \ \forall z \in S\big) \implies \big(\mathbb{P}(Z_1 = z) = \pi_z, \ \forall z \in S\big).$$

Proof Starting from the law of total probability

$$\mathbb{P}(Z_1 = z) = \sum_{\tilde{z} \in S} \mathbb{P}(Z_1 = z \mid Z_0 = \tilde{z}) \mathbb{P}(Z_0 = \tilde{z})$$

$$= \mathbb{P}(Z_1 = z \mid Z_0 = z) \mathbb{P}(Z_0 = z)$$

$$+ \sum_{k=1}^{N} \mathbb{P}(Z_1 = z \mid Z_0 = \tilde{z}^k) \mathbb{P}(Z_0 = \tilde{z}^k),$$

we show, using (7.3), that $\mathbb{P}(Z_1 = z)$ equals π_z if $\mathbb{P}(Z_0 = z) = \pi_z$ for all $z \in S$. Indeed, using (7.3) we have

$$\mathbb{P}(Z_1 = z) = \mathbb{P}(Z_1 = z \mid Z_0 = z)\pi_z + \sum_{k=1}^{N} \mathbb{P}(Z_1 = z \mid Z_0 = \tilde{z}^k)\pi_{\tilde{z}^k}$$

$$= \pi_z \mathbb{P}(Z_1 = z \mid Z_0 = z) + \pi_z \sum_{k=1}^{N} \mathbb{P}(Z_1 = \tilde{z}^k \mid Z_0 = z)$$

$$= \pi_z \left(\mathbb{P}(Z_1 = z \mid Z_0 = z) + \sum_{k=1}^{N} \mathbb{P}(Z_1 = \bar{z}^k \mid Z_0 = z) \right)$$

$$= \pi_z,$$

hence $(\pi_z)_{z \in S}$ is a stationary distribution for the chain $(Z_n)_{n \geqslant 0}$.   □

The stationary distribution $(\pi_z)_{z \in S}$ is known as the *Boltzmann distribution,* and is computed in the next proposition.

Proposition 7.3 *The probability distribution $(\pi_z)_{z \in S}$ defined as*

$$\pi_z := C_\beta \exp \left(\beta \sum_{l=0}^{N} z_l z_{l+1} \right), \qquad z \in S, \tag{7.4}$$

is the stationary and limiting distribution of $(Z_n)_{n \geqslant 0}$, where

$$C_\beta := \left(\sum_{z \in S} \exp \left(\beta \sum_{l=0}^{N} z_l z_{l+1} \right) \right)^{-1}$$

is a normalization constant and β is the inverse temperature *given in terms of p and q by*

$$\beta = \frac{1}{4} \log \frac{p}{q}, \quad i.e. \quad p = \frac{1}{1 + e^{-4\beta}}.$$

Proof Using Relation (7.5) in Lemma 7.4 below, we show that $(\pi_z)_{z \in S}$ defined in (7.4) satisfies (7.3). For all $z \in S$ we have

$$\pi_{\bar{z}^k} = C_\beta \exp \left(\beta \sum_{l=0}^{k-2} z_l z_{l+1} - \beta z_{k-1} z_k - \beta z_k z_{k+1} + \beta \sum_{l=k+1}^{N} z_l z_{l+1} \right)$$

$$= C_\beta \exp \left(-2\beta z_k (z_{k-1} + z_{k+1}) + \beta \sum_{l=0}^{N} z_l z_{l+1} \right)$$

$$= \pi_z e^{-2\beta z_k (z_{k-1} + z_{k+1})}$$

$$= \pi_z \left(\frac{q}{p} \right)^{z_k (z_{k-1} + z_{k+1})/2}$$

$$= \pi_z \frac{\mathbb{P}(Z_1 = \bar{z}^k \mid Z_0 = z)}{\mathbb{P}(Z_1 = z \mid Z_0 = \bar{z}^k)}, \qquad k = 1, 2, \ldots, N,$$

by (7.5) below, and the *inverse temperature* β is given by

$$\beta = \frac{1}{4} \log \frac{p}{q} = -\frac{1}{4} \log \left(\frac{1}{p} - 1\right),$$

i.e.

$$p = \frac{1}{1 + e^{-4\beta}}.$$

The constant C_β is chosen so that

$$1 = \sum_{z \in S} \pi_z = C_\beta \sum_{z \in S} \exp \left(\beta \sum_{l=0}^{N} z_l z_{l+1}\right),$$

i.e.

$$C_\beta = \left(\sum_{z \in S} \exp \left(\beta \sum_{l=0}^{N} z_l z_{l+1}\right)\right)^{-1}.$$

$\square$

More generally, the stationary distribution $(\pi_z)_{z \in S}$ can take the form

$$\pi_z := C_\beta e^{\beta H(z)} \qquad z \in S,$$

where

$$H(z) = \sum_{0 \leqslant i, j \leqslant N+1} J_{i,j} z_i z_j$$

is the *Hamiltonian* of the system, with

$$J_{i,j} = \mathbb{1}_{\{j=i+1\}}, \qquad 0 \leqslant i, j \leqslant N + 1,$$

in Proposition 7.3. More general Hamiltonians can be used to model long range interaction. We note that when the probability p of "agreeing" is larger than half, then the *temperature* $1/\beta$ is negative, whereas it is positive when $p < 1/2$.

In particular, when $-\beta < 0$ the configuration with the lowest probability $C_\beta e^{-(N+1)\beta}$ corresponds to a sequence $(z_k)_{0 \leqslant k \leqslant N+1}$ with alternating signs, while a constant spin sequence will have the highest probability $C_\beta e^{(N+1)\beta}$.

Conversely, when $-\beta > 0$ the configuration with the lowest probability $C_\beta e^{(N+1)\beta}$ corresponds to a constant spin sequence, while the highest probability $C_\beta e^{-(N+1)\beta}$ corresponds to a sequence $(z_k)_{0 \leqslant k \leqslant N+1}$ with alternating signs.

See Besag (1974) for the construction of a maximum pseudolikelihood estimate (MPLE) of β in the Ising model, Bhattacharya and Mukherjee (2018) for the consistency of this estimator, and Fig. 4 therein for an estimation of β with error bounds for a Facebook friendship-network in which spin values refer to gender.

The next lemma has been used in the proof of Proposition 7.3.

Lemma 7.4 *We have*

$$\frac{\mathbb{P}(Z_1 = \bar{z}^k \mid Z_0 = z)}{\mathbb{P}(Z_1 = z \mid Z_0 = \bar{z}^k)} = \left(\frac{q}{p}\right)^{z_k(z_{k-1}+z_{k+1})/2} , \qquad k = 1, 2, \ldots, N, \quad z \in S.$$

(7.5)

Proof By (7.2), we have

$$\frac{\mathbb{P}(Z_1 = \bar{z}^k \mid Z_0 = z)}{\mathbb{P}(Z_1 = z \mid Z_0 = \bar{z}^k)} = \frac{1 + (p/q)^{\bar{z}_k^k(\bar{z}_{k-1}^k + \bar{z}_{k+1}^k)/2}}{1 + (p/q)^{z_k(z_{k-1}+z_{k+1})/2}}$$

$$= \frac{1 + (p/q)^{-z_k(z_{k-1}+z_{k+1})/2}}{1 + (p/q)^{z_k(z_{k-1}+z_{k+1})/2}}$$

$$= \frac{q^{z_k(z_{k-1}+z_{k+1})/2}(1 + (q/p)^{z_k(z_{k-1}+z_{k+1})/2})}{q^{z_k(z_{k-1}+z_{k+1})/2} + p^{z_k(z_{k-1}+z_{k+1})/2}}$$

$$= \frac{(q/p)^{z_k(z_{k-1}+z_{k+1})/2}(p^{z_k(z_{k-1}+z_{k+1})/2} + q^{z_k(z_{k-1}+z_{k+1})/2})}{q^{z_k(z_{k-1}+z_{k+1})/2} + p^{z_k(z_{k-1}+z_{k+1})/2}}$$

$$= \left(\frac{q}{p}\right)^{z_k(z_{k-1}+z_{k+1})/2} , \qquad k = 1, 2, \ldots, N.$$

We could also have more directly used the relations

$$\frac{1 + x}{1 + 1/x} = x, \qquad x = \left(\frac{p}{q}\right)^{-\bar{z}^k(\bar{z}_{k-1}^k + \bar{z}_{k+1}^k)/2} > 0,$$

which imply

$$\frac{1 + (q/p)}{1 + (p/q)} = \frac{q}{p} \quad \text{and} \quad \frac{1 + (p/q)}{1 + (q/p)} = \frac{p}{q}.$$

□

7.4 Simulation Examples

In this section, we consider small scale simulation examples, although the real-life applications of the Ising model involve large values of N.

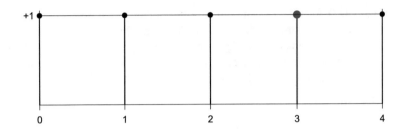

Fig. 7.3 Simulation with $N = 3$, $p = \sqrt{0.75} \approx 0.87$, $\beta = -0.47$, and $z_0 = z_4 = +1$

(i) Taking $N = 3$ and $z_0 = z_4 = +1$, i.e. $(z_0, z_1, z_2, z_3, z_4)$ takes the form $(+, \pm, \pm, \pm, +)$, we find the limiting distribution on the 8 configurations in

$$S = \{- - -, - - +, - + -, - - +, + - -, + - +, + + -, + + +\},$$

and we compute the value of C_β. Fig. 7.3 can be plotted using the ℝ code_17_page_204.R.

We have

$$\pi = \begin{bmatrix} \pi_{---} \\ \pi_{--+} \\ \pi_{-+-} \\ \pi_{-++} \\ \pi_{+--} \\ \pi_{+-+} \\ \pi_{++-} \\ \pi_{+++} \end{bmatrix} = \begin{bmatrix} C_\beta \\ C_\beta \\ C_\beta e^{-4\beta} \\ C_\beta \\ C_\beta \\ C_\beta \\ C_\beta \\ C_\beta e^{4\beta} \end{bmatrix} = C_\beta \begin{bmatrix} 1 \\ 1 \\ q/p \\ 1 \\ 1 \\ 1 \\ 1 \\ p/q \end{bmatrix} = \frac{1}{1 + 4pq} \begin{bmatrix} pq \\ pq \\ q^2 \\ pq \\ pq \\ pq \\ pq \\ p^2 \end{bmatrix} \begin{matrix} --- \\ --+ \\ -+- \\ -++ \\ +-- \\ +-+ \\ ++- \\ +++ \end{matrix}$$

with $q/p = \sqrt{3/4}/(1 - \sqrt{3/4}) = 6.46$, and from the relation

$$\pi_{---} + \pi_{--+} + \pi_{-+-} + \pi_{-++} + \pi_{+--} + \pi_{+-+} + \pi_{++-} + \pi_{+++} = 1,$$

we find

$$C_\beta = \frac{1}{e^{4\beta} + e^{-4\beta} + 6}$$

$$= \frac{1}{4\cosh^2(2\beta) + 4}$$

$$= \frac{pq}{q^2 + p^2 + 6pq}$$

$$= \frac{1}{6 + p/q + q/p}$$

$$= \frac{pq}{1 + 4pq}.$$

We note that when $p > 1/2$ the configuration "$++$" has the highest probability p^2, while "$-+-$" has the lowest probability q^2 in the long run, due to the presence of two "opinion leaders" $z_0 = +1$ and $z_4 = +1$ who will not change their minds.

We can also compute the probabilities of having more "+" than "-" in the long run, as

$$\pi_{-++} + \pi_{+-+} + \pi_{++-} + \pi_{+++} = \frac{(1 + 2q)p}{1 + 4pq},$$

see Fig. 7.4, while the probability of having more "-" than "+" is

$$\pi_{---} + \pi_{--+} + \pi_{-+-} + \pi_{+--} = \frac{(1 + 2p)q}{1 + 4pq}.$$

Clearly, the end result is influenced by the boundary conditions $z_0 = z_4 = +1$.

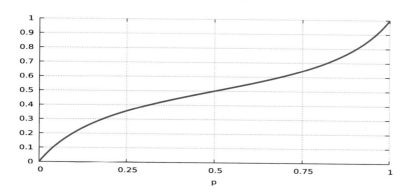

Fig. 7.4 Probability of a majority of "+" in the long run as a function of $p \in [0, 1]$

(ii) For another example, taking $z_0 = -1$ and $z_4 = +1$, we have

$$
\pi = \begin{bmatrix} \pi_{---} \\ \pi_{--+} \\ \pi_{-+-} \\ \pi_{-++} \\ \pi_{+--} \\ \pi_{+-+} \\ \pi_{++-} \\ \pi_{+++} \end{bmatrix} = \begin{bmatrix} C_\beta e^{2\beta} \\ C_\beta e^{2\beta} \\ C_\beta e^{-2\beta} \\ C_\beta e^{2\beta} \\ C_\beta e^{-2\beta} \\ C_\beta e^{-2\beta} \\ C_\beta e^{-2\beta} \\ C_\beta e^{2\beta} \end{bmatrix} = C_\beta \begin{bmatrix} \sqrt{p/q} \\ \sqrt{p/q} \\ \sqrt{q/p} \\ \sqrt{p/q} \\ \sqrt{q/p} \\ \sqrt{q/p} \\ \sqrt{q/p} \\ \sqrt{p/q} \end{bmatrix} = \frac{1}{4} \begin{bmatrix} p \\ p \\ q \\ p \\ q \\ q \\ q \\ p \end{bmatrix} \begin{array}{l} --- \\ --+ \\ -+- \\ -++ \\ +-- \\ +-+ \\ ++- \\ +++ \end{array}
$$

where

$$
C_\beta = \frac{1}{4\sqrt{p/q} + 4\sqrt{q/p}} = \frac{\sqrt{pq}}{4}.
$$

The probabilities of having more "+" than "-" in the long run are

$$
\pi_{-++} + \pi_{+-+} + \pi_{++-} + \pi_{+++} = \frac{1}{2}
$$

while the probability of having more "-" than "+" is also

$$
\pi_{---} + \pi_{--+} + \pi_{-+-} + \pi_{+--} = \frac{1}{2}.
$$

Notes

See e.g. Agapie and Höns (2007) for further reading, and § 7.7.2 of Barbu and Zhu (2020) for an application to image denoising.

Exercises

Exercise 7.1 We consider an ant moving randomly on the vertices of the 3-dimensional cube C_3 represented as

$$C_3 = \big\{(e_1, e_2, e_3) \; : \; e_1, e_2, e_3 \in \{0, 1\}\big\},$$

by choosing a new edge with probability $1/3$ at every time step.

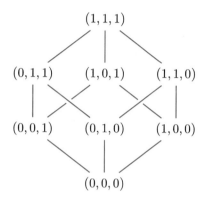

Using first step analysis, compute the mean time $h(r)$, $r = 0, 1, 2, 3$, until the ant reaches the vertex $(0, 0, 0)$ after starting from a vertex in the set S_r of vertices which are at distance $r = 0, 1, 2, 3$ from $(0, 0, 0)$, with $S_0 = \{(0, 0, 0)\}$ and $S_3 = \{(1, 1, 1)\}$.

Problem 7.2 We consider an ant moving randomly on the vertices of the d-dimensional (hyper)cube C_d represented as

$$C_d = \big\{(e_1, \ldots, e_d) \; : \; e_1, \ldots, e_d \in \{0, 1\}\big\},$$

by choosing a new edge with probability $1/d$ at every time step. We aim at computing the mean time $h(r)$ until the ant reaches the vertex $(0, \ldots, 0)$ after starting from a vertex in the set S_r of vertices which are at distance $r \in \{0, \ldots, d\}$ of $(1, \ldots, 1)$, with $S_0 = \{(1, \ldots, 1)\}$ and $S_d = \{(0, \ldots, 0)\}$.

(a) Give the value of $h(d)$.
(b) Find a relation between $h(0)$ and $h(1)$.
(c) Using first step analysis, find a relationship between $h(r)$, $h(r-1)$ and $h(r+1)$ for $r = 1, 2, \ldots, d-1$.
(d) Letting $f(r) := h(r+1) - h(r)$, $r = 0, 1, \ldots, d-1$, find a recurrence relation between $f(r)$ and $f(r-1)$ for $r = 1, 2, \ldots, d-1$.
(e) Find the value of $f(0)$ and solve the equation of Question (d) for $f(r)$, $r = 1, 2, \ldots, d$.

Hint. The solution of the equation

$$rf(r-1) = d + (d-r)f(r), \qquad r = 1, 2, \ldots, d,$$

with $f(0) = -1$ is given by

$$f(r) = -\frac{1}{\binom{d-1}{r}} \sum_{k=0}^{r} \binom{d}{k}, \qquad r = 0, 1, \ldots, d.$$

(f) Using a telescoping identity, find the value of $h(r)$ for $r = 0, 1, \ldots, d$.
(g) Give the values of $h(0)$, $h(1)$ and $h(2)$.
(h) Find the values of $h(r)$ for $r = 0, 1, \ldots, d$ in the following cases:

(i) $d = 1$,
(ii) $d = 2$,
(iii) $d = 3$.

Chapter 8
Search Engines

In this chapter we describe the PageRank™ and related ranking algorithms for search and meta search engines. This approach to ranking relies on the notions of limiting and stationary distributions presented in the previous chapters. We also apply the quantitative bounds on convergence to equilibrium discussed in Chap. 6.

8.1 Markovian Modeling of Ranking

PageRank™ algorithm. We consider the ranking of five web pages a, b, c, d, e which are linked according to the following sample graph.

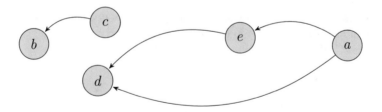

The algorithm works by constructing a self-improving random sequence $(X_n)_{n \geqslant 0}$ which is supposed to "converge" to the best possible search result. Given a search result $X_n = x \in S := \{a, b, c, d, e\}$, we choose the next search result X_{n+1} with the conditional probability

$$\mathbb{P}(X_{n+1} = y \mid X_n = x) = \frac{1}{n_x} \mathbb{1}_{\{x \to y\}}, \qquad x, y \in S,$$

where n_x denotes the number of outgoing links from x and "$x \to y$" means that x can lead to y in the graph. We also assume that "$x \to x$" is always true.

© The Author(s), under exclusive license to Springer Nature Switzerland AG 2024
N. Privault, *Discrete Stochastic Processes*, Springer Undergraduate
Mathematics Series, https://doi.org/10.1007/978-3-031-65820-4_8

The process $(X_n)_{n \geqslant 0}$ is a Markov chain with state space (a, b, c, d, e) and transition matrix

$$P = \begin{bmatrix} 0 & 0 & 0 & 1/2 & 1/2 \\ 0 & 1 & 0 & 0 & 0 \\ 0 & 1 & 0 & 0 & 0 \\ 0 & 0 & 0 & 1 & 0 \\ 0 & 0 & 0 & 1 & 0 \end{bmatrix}.$$

In addition, the chain $(X_n)_{n \geqslant 0}$ is clearly reducible, as can be seen from its graph:

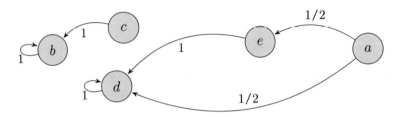

8.2 Limiting and Stationary Distributions

We note that the chain $(X_n)_{n \geqslant 0}$ admits a limiting distribution which is dependent of the initial state. Starting from state ⓐ, ⓓ or ⓔ, the limiting distribution is $(0, 0, 0, 1, 0)$, starting from state ⓑ or ⓒ, the limiting distribution is $(0, 1, 0, 0, 0)$, so that although the chain admits limiting distributions, it does *not* admit a limiting distribution independent of the initial state. More precisely, it can be checked that the powers P^n of the transition matrix P take the form

$$P^n = \begin{bmatrix} 0 & 0 & 0 & 1 & 0 \\ 0 & 1 & 0 & 0 & 0 \\ 0 & 1 & 0 & 0 & 0 \\ 0 & 0 & 0 & 1 & 0 \\ 0 & 0 & 0 & 1 & 0 \end{bmatrix} \quad \text{for all } n \geqslant 2, \text{ hence } \lim_{n \to \infty} P^n = \begin{bmatrix} 0 & 0 & 0 & 1 & 0 \\ 0 & 1 & 0 & 0 & 0 \\ 0 & 1 & 0 & 0 & 0 \\ 0 & 0 & 0 & 1 & 0 \\ 0 & 0 & 0 & 1 & 0 \end{bmatrix}.$$

The following proposition shows that the stationary distribution is not unique here because the chain is reducible.

Proposition 8.1 *Any probability distribution of the form*

$$\pi = [\pi_a, \pi_b, \pi_c, \pi_d, \pi_e] = [0, p, 0, 1 - p, 0],$$

with $p \in [0, 1]$, is a stationary distribution for the chain with matrix P.

Proof The equation $\pi = \pi P$ is satisfied by any probability distribution of the form

$$\pi = [\pi_a, \pi_b, \pi_c, \pi_d, \pi_e] = [0, p, 0, 1 - p, 0],$$

with $p \in [0, 1]$. $\square$

Clearly, in the long run the chain $(X_k)_{k \in \mathbb{N}}$ will converge to state $\widehat{b}$ if it starts from $\widehat{c}$ or $\widehat{b}$, and it will converge to state $\widehat{d}$ if it starts from $\widehat{a}$, $\widehat{d}$, or $\widehat{e}$. However, this does not allow us to compare the states $\widehat{b}$ and $\widehat{d}$. This issue is addressed in the next section.

8.3 Matrix Perturbation

In PageRank™-type algorithms, one typically chooses to perturb the transition matrix P into the new matrix

$$P(\varepsilon) := \frac{\varepsilon}{n} \begin{bmatrix} 1 & 1 & 1 & 1 & 1 \\ 1 & 1 & 1 & 1 & 1 \\ 1 & 1 & 1 & 1 & 1 \\ 1 & 1 & 1 & 1 & 1 \\ 1 & 1 & 1 & 1 & 1 \end{bmatrix} + (1 - \varepsilon)P$$

$$= \begin{bmatrix} \dfrac{\varepsilon}{5} & \dfrac{\varepsilon}{5} & \dfrac{\varepsilon}{5} & \dfrac{5 - 3\varepsilon}{10} & \dfrac{5 - 3\varepsilon}{10} \\[2mm] \dfrac{\varepsilon}{5} & \dfrac{5 - 4\varepsilon}{5} & \dfrac{\varepsilon}{5} & \dfrac{\varepsilon}{5} & \dfrac{\varepsilon}{5} \\[2mm] \dfrac{\varepsilon}{5} & \dfrac{5 - 4\varepsilon}{5} & \dfrac{\varepsilon}{5} & \dfrac{\varepsilon}{5} & \dfrac{\varepsilon}{5} \\[2mm] \dfrac{\varepsilon}{5} & \dfrac{\varepsilon}{5} & \dfrac{\varepsilon}{5} & \dfrac{5 - 4\varepsilon}{5} & \dfrac{\varepsilon}{5} \\[2mm] \dfrac{\varepsilon}{5} & \dfrac{\varepsilon}{5} & \dfrac{\varepsilon}{5} & \dfrac{5 - 4\varepsilon}{5} & \dfrac{\varepsilon}{5} \end{bmatrix},$$

with $n = 5$ here, and $\varepsilon \in (0, 1)$, with $1 - \varepsilon$ referred to as the *damping factor*.

We note that $P(\varepsilon)$ is a Markov transition matrix, and that the corresponding chain $\left(X_n^{(\varepsilon)}\right)_{n \geqslant 1}$ is irreducible and aperiodic. Indeed, all rows in the matrix $P(\varepsilon)$ clearly add up to 1, so $P(\varepsilon)$ is a Markov transition matrix. On the other hand, all states become accessible from each other so that the new chain is irreducible and all states have period 1.

Since the chain is irreducible, aperiodic and has a finite state space, we know by Corollary 6.7 that it admits a unique limiting and stationary distribution $\pi(\varepsilon)$. For example, taking with $\varepsilon = 0.1$ and $n = 200$, we have

$$
P(\varepsilon)^n = \left(\frac{\varepsilon}{5} \begin{bmatrix} 1 & 1 & 1 & 1 & 1 \\ 1 & 1 & 1 & 1 & 1 \\ 1 & 1 & 1 & 1 & 1 \\ 1 & 1 & 1 & 1 & 1 \\ 1 & 1 & 1 & 1 & 1 \end{bmatrix} + (1-\varepsilon) \begin{bmatrix} 0 & 0 & 0 & 0.5 & 0.5 \\ 0 & 1 & 0 & 0 & 0 \\ 0 & 1 & 0 & 0 & 0 \\ 0 & 0 & 0 & 1 & 0 \\ 0 & 0 & 0 & 1 & 0 \end{bmatrix} \right)^{200}
$$

$$
= \begin{bmatrix} 0.02 & 0.38 & 0.02 & 0.551 & 0.029 \\ 0.02 & 0.38 & 0.02 & 0.551 & 0.029 \\ 0.02 & 0.38 & 0.02 & 0.551 & 0.029 \\ 0.02 & 0.38 & 0.02 & 0.551 & 0.029 \\ 0.02 & 0.38 & 0.02 & 0.551 & 0.029 \end{bmatrix},
$$

which can be obtained in Mathematica via the command

MatrixPower[(0.1/5)*[[1,1,1,1,1],[1,1,1,1,1],[1,1,1,1,1],[1,1,1,1,1],[1,1,1,1,1]]
+0.9*[[0,0,0,0.5,0.5],[0,1,0,0,0],[0,1,0,0,0],[0,0,0,1,0],[0,0,0,1,0]],200]

with $\varepsilon = 0.1$. Since the chain is irreducible, aperiodic and has a finite state space, we know by Corollary 6.7 that the limiting distribution $\pi(\varepsilon)$ is also the unique stationary distribution of the chain, which can be determined by solving the equation $\pi(\varepsilon) = \pi(\varepsilon)P(\varepsilon)$, i.e.

$$
\pi(\varepsilon) = \pi(\varepsilon)P(\varepsilon)
$$

$$
= \frac{\varepsilon}{n}\pi(\varepsilon) \begin{bmatrix} 1 & 1 & 1 & 1 & 1 \\ 1 & 1 & 1 & 1 & 1 \\ 1 & 1 & 1 & 1 & 1 \\ 1 & 1 & 1 & 1 & 1 \\ 1 & 1 & 1 & 1 & 1 \end{bmatrix} + (1-\varepsilon)\pi(\varepsilon)P
$$

$$
= \left[\frac{\varepsilon}{5}, \frac{\varepsilon}{5}, \frac{\varepsilon}{5}, \frac{\varepsilon}{5}, \frac{\varepsilon}{5} \right] + (1-\varepsilon)\pi(\varepsilon)P.
$$

From the above calculation, we check that all probabilities in $\pi(\varepsilon)$ are greater than $\varepsilon/5$.

Proposition 8.2 *The limiting and stationary distribution of $P(\varepsilon)$ is given by*

$$
\begin{cases} \pi_a(\varepsilon) = \dfrac{\varepsilon}{5}, & \pi_b(\varepsilon) = \dfrac{2-\varepsilon}{5}, & \pi_c(\varepsilon) = \dfrac{\varepsilon}{5}, \\[3mm] \pi_d(\varepsilon) = \dfrac{(2-\varepsilon)(3-\varepsilon)}{10}, & \pi_e(\varepsilon) = \dfrac{(3-\varepsilon)\varepsilon}{10}. \end{cases} \tag{8.1}
$$

Proof The equation

$$\pi(\varepsilon) = \left[\frac{\varepsilon}{5}, \frac{\varepsilon}{5}, \frac{\varepsilon}{5}, \frac{\varepsilon}{5}, \frac{\varepsilon}{5}\right] + (1 - \varepsilon)\pi(\varepsilon)P$$

reads

$$[\pi_a(\varepsilon), \pi_b(\varepsilon), \pi_c(\varepsilon), \pi_d(\varepsilon), \pi_e(\varepsilon)] = \left[\frac{\varepsilon}{5}, \frac{\varepsilon}{5}, \frac{\varepsilon}{5}, \frac{\varepsilon}{5}, \frac{\varepsilon}{5}\right]$$

$$+ (1 - \varepsilon)\pi(\varepsilon) \begin{bmatrix} 0 & 0 & 0 & 1/2 & 1/2 \\ 0 & 1 & 0 & 0 & 0 \\ 0 & 1 & 0 & 0 & 0 \\ 0 & 0 & 0 & 1 & 0 \\ 0 & 0 & 0 & 1 & 0 \end{bmatrix},$$

which yields (8.1). □

Note that the stationary distribution π can also be obtained as $\pi = \eta^\top$, where η is the (normalized) eigenvector of eigenvalue 1 of the transposed transition matrix $P^\top$, i.e. such that $\eta = P^\top \eta$, that can be obtained in Mathematica via the command

Eigenvectors[(epsilon/5)*[[1,1,1,1,1],[1,1,1,1,1],[1,1,1,1,1],[1,1,1,1,1],[1,1,1,1,1]]
    +(1-epsilon)*0,0,0,0,0,0,1,1,0,0,0,0,0,0,0,0.5,0,0,1,1,0.5,0,0,0,0],

see also Bryan and Leise (2006).

8.4 State Ranking

We are now ready to provide a ranking of the states $\{a, b, c, d, e\}$ based on the limiting and stationary distribution $\pi(\varepsilon)$ which is plotted as a function of $\varepsilon \in [0, 1]$ in Fig. 8.1.

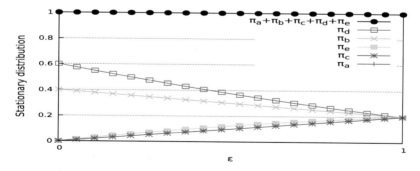

Fig. 8.1 Stationary distribution as a function of $\varepsilon \in [0, 1]$

We note that

$$\pi_a(\varepsilon) = \pi_c(\varepsilon) < \pi_e(\varepsilon) < \pi_b(\varepsilon) < \pi_d(\varepsilon), \qquad \varepsilon \in (0, 1],$$

hence we will rank the states as

Rank	State
1	d
2	b
3	e
4	$a \simeq c$

based on the idea that the most visited states should rank higher.

Convergence Analysis

We note that, proceeding similarly to (6.18), Assumption (C) p. 169, i.e.

$$[P(\varepsilon)]_{i,j} \geqslant \theta(\varepsilon)\pi_j(\varepsilon), \qquad i, j \in \mathsf{S},$$

reads, using componentwise ordering,

$$
\begin{bmatrix}
\dfrac{\varepsilon}{5} & \dfrac{\varepsilon}{5} & \dfrac{\varepsilon}{5} & \dfrac{5-3\varepsilon}{10} & \dfrac{5-3\varepsilon}{10} \\[2mm]
\dfrac{\varepsilon}{5} & \dfrac{5-4\varepsilon}{5} & \dfrac{\varepsilon}{5} & \dfrac{\varepsilon}{5} & \dfrac{\varepsilon}{5} \\[2mm]
\dfrac{\varepsilon}{5} & \dfrac{5-4\varepsilon}{5} & \dfrac{\varepsilon}{5} & \dfrac{\varepsilon}{5} & \dfrac{\varepsilon}{5} \\[2mm]
\dfrac{\varepsilon}{5} & \dfrac{\varepsilon}{5} & \dfrac{\varepsilon}{5} & \dfrac{5-4\varepsilon}{5} & \dfrac{\varepsilon}{5} \\[2mm]
\dfrac{\varepsilon}{5} & \dfrac{\varepsilon}{5} & \dfrac{\varepsilon}{5} & \dfrac{5-4\varepsilon}{5} & \dfrac{\varepsilon}{5} \\[2mm]
\dfrac{\varepsilon}{5} & \dfrac{\varepsilon}{5} & \dfrac{\varepsilon}{5} & \dfrac{\varepsilon}{5} & \dfrac{\varepsilon}{5}
\end{bmatrix}
$$

$$
\geqslant \theta(\varepsilon) \times
\begin{bmatrix}
\dfrac{\varepsilon}{5} & \dfrac{2-\varepsilon}{5} & \dfrac{\varepsilon}{5} & \dfrac{(2-\varepsilon)(3-\varepsilon)}{10} & \dfrac{(3-\varepsilon)\varepsilon}{10} \\[2mm]
\dfrac{\varepsilon}{5} & \dfrac{2-\varepsilon}{5} & \dfrac{\varepsilon}{5} & \dfrac{(2-\varepsilon)(3-\varepsilon)}{10} & \dfrac{(3-\varepsilon)\varepsilon}{10} \\[2mm]
\dfrac{\varepsilon}{5} & \dfrac{2-\varepsilon}{5} & \dfrac{\varepsilon}{5} & \dfrac{(2-\varepsilon)(3-\varepsilon)}{10} & \dfrac{(3-\varepsilon)\varepsilon}{10} \\[2mm]
\dfrac{\varepsilon}{5} & \dfrac{2-\varepsilon}{5} & \dfrac{\varepsilon}{5} & \dfrac{(2-\varepsilon)(3-\varepsilon)}{10} & \dfrac{(3-\varepsilon)\varepsilon}{10} \\[2mm]
\dfrac{\varepsilon}{5} & \dfrac{2-\varepsilon}{5} & \dfrac{\varepsilon}{5} & \dfrac{(2-\varepsilon)(3-\varepsilon)}{10} & \dfrac{(3-\varepsilon)\varepsilon}{10} \\[2mm]
\dfrac{\varepsilon}{5} & \dfrac{2-\varepsilon}{5} & \dfrac{\varepsilon}{5} & \dfrac{10}{10} & \dfrac{10}{10}
\end{bmatrix},
$$

or equivalently

$$
\begin{bmatrix}
1 & \dfrac{\varepsilon}{2-\varepsilon} & 1 & \dfrac{5-3\varepsilon}{(2-\varepsilon)(3-\varepsilon)} & \dfrac{5-3\varepsilon}{(3-\varepsilon)\varepsilon} \\[2ex]
1 & \dfrac{5-4\varepsilon}{2-\varepsilon} & 1 & \dfrac{2\varepsilon}{(2-\varepsilon)(3-\varepsilon)} & \dfrac{2\varepsilon}{(3-\varepsilon)\varepsilon} \\[2ex]
1 & \dfrac{5-4\varepsilon}{2-\varepsilon} & 1 & \dfrac{2\varepsilon}{(2-\varepsilon)(3-\varepsilon)} & \dfrac{2\varepsilon}{(3-\varepsilon)\varepsilon} \\[2ex]
1 & \dfrac{\varepsilon}{2-\varepsilon} & 1 & \dfrac{10-8\varepsilon}{(2-\varepsilon)(3-\varepsilon)} & \dfrac{2\varepsilon}{(3-\varepsilon)\varepsilon} \\[2ex]
1 & \dfrac{\varepsilon}{2-\varepsilon} & 1 & \dfrac{10-8\varepsilon}{(2-\varepsilon)(3-\varepsilon)} & \dfrac{2\varepsilon}{(3-\varepsilon)\varepsilon}
\end{bmatrix}
$$

$$
\geqslant
\begin{bmatrix}
\theta(\varepsilon) & \theta(\varepsilon) & \theta(\varepsilon) & \theta(\varepsilon) & \theta(\varepsilon) \\
\theta(\varepsilon) & \theta(\varepsilon) & \theta(\varepsilon) & \theta(\varepsilon) & \theta(\varepsilon) \\
\theta(\varepsilon) & \theta(\varepsilon) & \theta(\varepsilon) & \theta(\varepsilon) & \theta(\varepsilon) \\
\theta(\varepsilon) & \theta(\varepsilon) & \theta(\varepsilon) & \theta(\varepsilon) & \theta(\varepsilon) \\
\theta(\varepsilon) & \theta(\varepsilon) & \theta(\varepsilon) & \theta(\varepsilon) & \theta(\varepsilon)
\end{bmatrix},
$$

which is satisfied for

$$
\theta(\varepsilon) = \frac{\varepsilon}{5\pi_d(\varepsilon)} = \frac{2\varepsilon}{(2-\varepsilon)(3-\varepsilon)} = 1 - \frac{(6-\varepsilon)(1-\varepsilon)}{(2-\varepsilon)(3-\varepsilon)}, \qquad \varepsilon \in (0,1).
$$

See Fig. 8.2. From Proposition 6.21, the convergence to the stationary distribution π occurs with speed at least equal to

$$
\begin{aligned}
d(n) &:= \operatorname*{Max}_{k=1,2,\ldots,N} \left\| [P^n]_{k,\cdot} - \pi \right\|_{\mathrm{TV}} \\
&\leqslant (1 - \theta(\varepsilon))^n \\
&= \left(\frac{(6-\varepsilon)(1-\varepsilon)}{(2-\varepsilon)(3-\varepsilon)} \right)^n, \qquad n \geqslant 1,
\end{aligned}
$$

see also Bryan and Leise (2006) and Exercise 6.7.

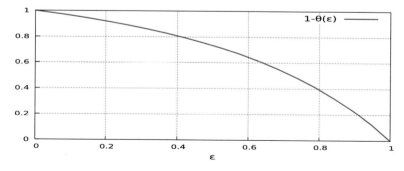

Fig. 8.2 Graph of $1 - \theta(\varepsilon)$ as a function of $\varepsilon \in [0,1]$

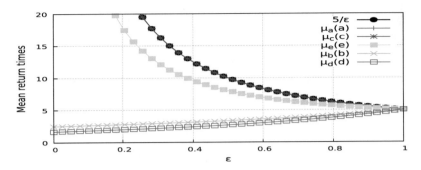

Fig. 8.3 Mean return times as functions of $\varepsilon \in [0, 1]$

We note that as ε tends to zero, Fig. 8.1 allows us to select a stationary distribution

$$\pi := \lim_{\varepsilon \to 0} \pi(\varepsilon) = \lim_{\varepsilon \to 0} [\pi_a(\varepsilon), \pi_b(\varepsilon), \pi_c(\varepsilon), \pi_d(\varepsilon), \pi_e(\varepsilon)] = [0, 2/5, 0, 3/5, 0]$$

which is consistent with Proposition 8.1.

Mean Return Times Analysis

By Theorem 6.6 and Proposition 8.2 we obtain the mean return times for $P(\varepsilon)$, and we note that they all remain below $5/\epsilon$. We have

$$\begin{cases} \mu_a(a) = \dfrac{5}{\varepsilon}, \quad \mu_b(b) = \dfrac{5}{2 - \varepsilon}, \quad \mu_c(c) = \dfrac{5}{\varepsilon}, \\[4mm] \mu_d(d) = \dfrac{10}{(2 - \varepsilon)(3 - \varepsilon)}, \quad \mu_e(e) = \dfrac{10}{(3 - \varepsilon)\varepsilon}. \end{cases}$$

We have $\lim_{\varepsilon \to 0} \mu_a(a) = \lim_{\varepsilon} \mu_c(c) = \lim_{\varepsilon} \mu_e(e) = +\infty$, and

$$\lim_{\varepsilon \to 0} \mu_b(b) = \frac{5}{2}, \qquad \lim_{\varepsilon \to 0} \mu_d(d) = \frac{5}{3},$$

which do not recover the values $\mu_b(b) = \mu_d(d) = 1$ in case $\varepsilon = 0$. In the graph of Fig. 8.3 the mean return times are plotted as a function of $\varepsilon \in [0, 1]$. A commonly used value in the literature is $\varepsilon = 1/7 \simeq 0.14$.

We note that the ranking of states is clearer for smaller values of ε. In particular ε cannot be chosen too large, for example taking $\varepsilon = 1$ makes all mean return times equal and corresponds to a uniform stationary distribution. However, the mean return times can be higher and hence the simulations can take longer for small values of ε. This type of algorithm can contribute to the creation of *link farms* as it tends to give higher rankings to the pages that have the most backlinks. The following ℝ code provides a realization of the Markov chain $(X_n)_{n \geq \mathbb{N}}$, see Fig. 8.4.

```
library(igraph); library(markovchain)
P<-matrix(c(0,0,0,0.5,0.5,0,1,0,0,0,0,1,0,0,0,0,0,0,1,0,0,0,0,1,0),nrow=5, byrow=TRUE)
MC <-new("markovchain", transitionMatrix=P, states=c("a","b","c","d","e"))
page_rank(graph,damping=1-epsilon)
```

with the output

```
$vector
    a  b  c  d  e
0.00600 0.39400 0.00600 0.58509 0.00891
```

This output can be recovered by calculation of the relevant stationary distribution, as follows.

```
I <- matrix(data=1,nrow=5,ncol=5); Pe<-epsilon*I/5+(1-epsilon)*P
MCe <-new("markovchain", transitionMatrix=Pe, states=c("a","b","c","d","e"))
```

with the output

```
steadyStates(object = MCe)
      a  b  c  d  e
[1,] 0.006 0.394 0.006 0.58509 0.00891
```

Fig. 8.4 Markovchain package output

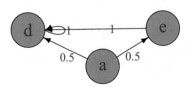

8.5   Meta Search Engines

In this section we consider a *meta search engine* which attempts to provide a single optimized ranking of search results $\{a, b, c, d, e\}$ based on the outputs of *4* different search engines denoted S_1, S_2, S_3, S_4, a technique known as *rank aggregation*, see Schalekamp and van Zuylen (2009) for further reading. Precisely, we consider four search engines S_1, S_2, S_3, S_4 and five possible search results a, b, c, d, e which have been respectively ranked as

Rank	S_1	S_2	S_3	S_4
1	b	c	d	e
2	c	b	e	a
3	d	d	a	d
4	a	e	b	b
5	e	a	c	c

by S_1, S_2, S_3, S_4.

Definition 8.3 (Partial Ordering) A state $y \in \{a, b, c, d, e\}$ is said to be better ranked than another state $x \in \{a, b, c, d, e\}$, and we write $x \preceq y$ and $y \succeq x$ if y ranks higher than x in *at least three* of the four search rankings.

We also write "$x \npreceq y$" when neither "$x \preceq y$" nor "$x \succeq y$" is satisfied. A ranking table for the order $\preceq$ can be completed as follows, using "$x \preceq y$" or "$x \npreceq y$" at the position (x, y).

$\preceq$	a	b	c	d	e
a	$=$	$\npreceq$	$\npreceq$	$\preceq$	$\preceq$
b	$\npreceq$	$=$	$\succeq$	$\npreceq$	$\npreceq$
c	$\npreceq$	$\preceq$	$=$	$\npreceq$	$\npreceq$
d	$\succeq$	$\npreceq$	$\npreceq$	$=$	$\succeq$
e	$\succeq$	$\npreceq$	$\npreceq$	$\preceq$	$=$

The diagonal entries, which are marked with "$=$", are not relevant here.

The meta search engine works by constructing a self-improving random sequence $(X_n)_{n \geqslant 0}$ on a state space $S = \{a, b, c, d, e\}$ of websites, which is supposed to "converge" to the best possible search result based on the data of the four rankings.

Given a search result $X_n = x$ we choose the next search result X_{n+1} by assigning probability $1/5$ to each of the search results that are *better ranked* than x. If no search result is better than x, then we keep $X_{n+1} = x$.

The process $(X_n)_{n \geqslant 0}$ is a Markov chain with state space (a, b, c, d, e) and transition matrix

$$
P = \begin{bmatrix}
3/5 & 0 & 0 & 1/5 & 1/5 \\
0 & 1 & 0 & 0 & 0 \\
0 & 1/5 & 4/5 & 0 & 0 \\
0 & 0 & 0 & 1 & 0 \\
0 & 0 & 0 & 1/5 & 4/5
\end{bmatrix}.
$$

In addition, the chain $(X_n)_{n \geqslant 0}$ is clearly reducible, as can be seen from its graph:

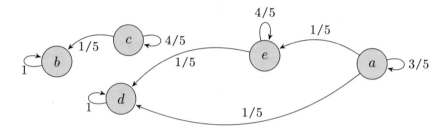

Limiting and Stationary Distributions

We note that the chain $(X_n)_{n \geqslant 0}$ admits a limiting distribution which is dependent on the initial state. Starting from states (a), (d) or (e), the limiting distribution is $(0, 0, 0, 1, 0)$, starting from states (b) or (c), the limiting distribution is $(0, 1, 0, 0, 0)$. More precisely, we check that the power P^n of order $n \geqslant 1$ of the transition matrix P takes the form

$$
P^n = \begin{bmatrix}
(3/5)^n & 0 & 0 & 1 - (4/5)^n & (4/5)^n - (3/5)^n \\
0 & 1 & 0 & 0 & 0 \\
0 & 1 - (4/5)^n & (4/5)^n & 0 & 0 \\
0 & 0 & 0 & 1 & 0 \\
0 & 0 & 0 & 1 - (4/5)^n & (4/5)^n
\end{bmatrix}
$$

hence

$$
\lim_{n \to \infty} P^n = \begin{bmatrix}
0 & 0 & 0 & 1 & 0 \\
0 & 1 & 0 & 0 & 0 \\
0 & 1 & 0 & 0 & 0 \\
0 & 0 & 0 & 1 & 0 \\
0 & 0 & 0 & 1 & 0
\end{bmatrix}.
$$

The following proposition shows that the stationary distribution is not unique here because the chain is reducible.

Proposition 8.4 *Any probability distribution of the form*

$$\pi(\varepsilon) = [\pi_a(\varepsilon), \pi_b(\varepsilon), \pi_c(\varepsilon), \pi_d(\varepsilon), \pi_e(\varepsilon)] = [0, p, 0, 1 - p, 0],$$

with $p \in [0, 1]$, is a stationary distribution for the chain with matrix P.

Proof The stationary distribution(s) of the chain $(X_n)_{n \geq 0}$ can be found by solving the equation

$$\pi(\varepsilon) = \pi(\varepsilon)P$$

which reads

$$\pi(\varepsilon) = [\pi_a(\varepsilon), \pi_b(\varepsilon), \pi_c(\varepsilon), \pi_d(\varepsilon), \pi_e(\varepsilon)]$$

$$= \pi \begin{bmatrix} 3/5 & 0 & 0 & 1/5 & 1/5 \\ 0 & 1 & 0 & 0 & 0 \\ 0 & 1/5 & 4/5 & 0 & 0 \\ 0 & 0 & 0 & 1 & 0 \\ 0 & 0 & 0 & 1/5 & 4/5 \end{bmatrix}$$

$$= \left[3\frac{\pi_a(\varepsilon)}{5}, \pi_b(\varepsilon) + \frac{\pi_c(\varepsilon)}{5}, 4\frac{\pi_c(\varepsilon)}{5}, \frac{\pi_a(\varepsilon)}{5} + \pi_d(\varepsilon) \right.$$

$$\left. + \frac{\pi_e(\varepsilon)}{5}, \frac{\pi_a(\varepsilon)}{5} + 4\frac{\pi_e(\varepsilon)}{5} \right],$$

i.e.

$$[0, 0, 0, 0, 0] = \left[-2\frac{\pi_a(\varepsilon)}{5}, \frac{\pi_c(\varepsilon)}{5}, -\frac{\pi_c(\varepsilon)}{5}, \frac{\pi_a(\varepsilon)}{5} + \frac{\pi_e(\varepsilon)}{5}, \frac{\pi_a(\varepsilon)}{5} - \frac{\pi_e(\varepsilon)}{5} \right],$$

or

$$\begin{cases} \pi_a(\varepsilon) = 0, \\ \pi_c(\varepsilon) = 0, \\ \pi_e(\varepsilon) = 0. \end{cases}$$

Therefore, based on the normalization condition

$$\pi_a(\varepsilon) + \pi_b(\varepsilon) + \pi_c(\varepsilon) + \pi_d(\varepsilon) + \pi_e(\varepsilon) = 1,$$

any probability distribution of the form

$$\pi(\varepsilon) = [\pi_a(\varepsilon), \pi_b(\varepsilon), \pi_c(\varepsilon), \pi_d(\varepsilon), \pi_e(\varepsilon)] = [0, p, 0, 1 - p, 0],$$

with $p \in [0, 1]$, will be a stationary distribution for the chain with matrix P.   □

Clearly, in the long run the chain $(X_k)_{k \in \mathbb{N}}$ will converge to state ⓑ if it starts from ⓒ or ⓑ, and it will converge to state ⓓ if it starts from ⓐ, ⓓ, or ⓔ. However, this does not allow us to compare the states ⓑ and ⓓ. This issue is addressed in the next section.

Matrix Perturbation

In PageRank™-type algorithms, one typically chooses to perturb the transition matrix P into the new matrix

$$P(\varepsilon) := \frac{\varepsilon}{n} \begin{bmatrix} 1 & 1 & 1 & 1 & 1 \\ 1 & 1 & 1 & 1 & 1 \\ 1 & 1 & 1 & 1 & 1 \\ 1 & 1 & 1 & 1 & 1 \\ 1 & 1 & 1 & 1 & 1 \end{bmatrix} + (1 - \varepsilon)P,$$

with $n = 5$ here, and $\varepsilon \in (0, 1)$.

We note that $P(\varepsilon)$ is a Markov transition matrix, and that the corresponding chain $\left(X_n^{(\varepsilon)}\right)_{n \geqslant 1}$ is irreducible and aperiodic. Indeed, all rows in the matrix $P(\varepsilon)$ clearly add up to 1, so $P(\varepsilon)$ is a Markov transition matrix. On the other hand, all states become accessible from each other so that the new chain is irreducible and all states have period 1.

Since the chain is irreducible, aperiodic and has a finite state space, we know by Corollary 6.7 that it admits a unique stationary distribution $\pi(\varepsilon)$. The equation $\pi(\varepsilon) = \pi(\varepsilon) P_\epsilon$ reads

$$\pi(\varepsilon) = \pi(\varepsilon) P_\epsilon$$

$$= \frac{\varepsilon}{n} \pi(\varepsilon) \begin{bmatrix} 1 & 1 & 1 & 1 & 1 \\ 1 & 1 & 1 & 1 & 1 \\ 1 & 1 & 1 & 1 & 1 \\ 1 & 1 & 1 & 1 & 1 \\ 1 & 1 & 1 & 1 & 1 \end{bmatrix} + (1 - \varepsilon)\pi(\varepsilon) P$$

$$= \left[\frac{\varepsilon}{5}, \frac{\varepsilon}{5}, \frac{\varepsilon}{5}, \frac{\varepsilon}{5}, \frac{\varepsilon}{5} \right] + (1 - \varepsilon)\pi(\varepsilon) P.$$

From the above calculation, we check that all probabilities in $\pi(\varepsilon)$ are greater than $\varepsilon/5$.

Proposition 8.5 *The limiting and stationary distribution of $P(\varepsilon)$ is given by*

$$
\begin{cases}
\pi_a(\varepsilon) = \dfrac{\varepsilon}{2+3\varepsilon}, & \pi_b(\varepsilon) = \dfrac{2+3\varepsilon}{5(1+4\varepsilon)}, & \pi_c(\varepsilon) = \dfrac{\varepsilon}{1+4\varepsilon}, \\[3mm]
\pi_d(\varepsilon) = \dfrac{3+2\varepsilon}{5(1+4\varepsilon)}, & \pi_e(\varepsilon) = \dfrac{\varepsilon(3+2\varepsilon)}{(1+4\varepsilon)(2+3\varepsilon)}.
\end{cases} \tag{8.2}
$$

Proof The equation

$$
\pi(\varepsilon) = \left[\frac{\varepsilon}{5}, \frac{\varepsilon}{5}, \frac{\varepsilon}{5}, \frac{\varepsilon}{5}, \frac{\varepsilon}{5} \right] + (1-\varepsilon)\pi(\varepsilon)P
$$

reads

$$
[\pi_a(\varepsilon), \pi_b(\varepsilon), \pi_c(\varepsilon), \pi_d(\varepsilon), \pi_e(\varepsilon)] = \left[\frac{\varepsilon}{5}, \frac{\varepsilon}{5}, \frac{\varepsilon}{5}, \frac{\varepsilon}{5}, \frac{\varepsilon}{5} \right]
$$

$$
+ (1-\varepsilon)\pi(\varepsilon)
\begin{bmatrix}
3/5 & 0 & 0 & 1/5 & 1/5 \\
0 & 1 & 0 & 0 & 0 \\
0 & 1/5 & 4/5 & 0 & 0 \\
0 & 0 & 0 & 1 & 0 \\
0 & 0 & 0 & 1/5 & 4/5
\end{bmatrix}
$$

$$
= \left[\frac{\varepsilon}{5}, \frac{\varepsilon}{5}, \frac{\varepsilon}{5}, \frac{\varepsilon}{5}, \frac{\varepsilon}{5} \right]
$$

$$
+ (1-\varepsilon)\left[3\frac{\pi_a(\varepsilon)}{5}, \pi_b(\varepsilon) + \frac{\pi_c(\varepsilon)}{5}, 4\frac{\pi_c(\varepsilon)}{5}, \frac{\pi_a(\varepsilon)}{5} + \pi_d(\varepsilon) \right.
$$

$$
\left. + \frac{\pi_e(\varepsilon)}{5}, \frac{\pi_a(\varepsilon)}{5} + 4\frac{\pi_e(\varepsilon)}{5} \right],
$$

i.e.

$[0, 0, 0, 0, 0]$

$$
= [\varepsilon + 3(1-\varepsilon)\pi_a(\varepsilon) - 5\pi_a(\varepsilon), \varepsilon + 5(1-\varepsilon)\pi_b(\varepsilon) - 5\pi_b(\varepsilon) + (1-\varepsilon)\pi_c(\varepsilon),
$$

$$
\varepsilon + 4(1-\varepsilon)\pi_c(\varepsilon) - 5\pi_c(\varepsilon), \varepsilon + (1-\varepsilon)\pi_a(\varepsilon) + 5(1-\varepsilon)\pi_d(\varepsilon)
$$

$$
-5\pi_d(\varepsilon) + (1-\varepsilon)\pi_e(\varepsilon), \varepsilon + (1-\varepsilon)\pi_a(\varepsilon) + 4(1-\varepsilon)\pi_e(\varepsilon) - 5\pi_e(\varepsilon)],
$$

i.e.

$$\begin{cases} \varepsilon - (2 + 3\varepsilon)\pi_a(\varepsilon) = 0 \\ \varepsilon - 5\varepsilon\pi_b(\varepsilon) + (1 - \varepsilon)\pi_c(\varepsilon) = 0 \\ \varepsilon - \pi_c(\varepsilon)(1 + 4\varepsilon) = 0 \\ \varepsilon + (1 - \varepsilon)\pi_a(\varepsilon) - 5\varepsilon\pi_d(\varepsilon) + (1 - \varepsilon)\pi_e(\varepsilon) = 0 \\ \varepsilon + (1 - \varepsilon)\pi_a(\varepsilon) - (1 + 4\varepsilon)\pi_e(\varepsilon) = 0, \end{cases}$$

which yields (8.2). $\qquad\qquad\qquad\qquad\qquad\qquad\qquad\qquad\qquad\qquad\qquad$ $\square$

As in Proposition 8.2, the stationary distribution π can be obtained as the transposed vector $\pi = \eta^\top$, where η is the (normalized) eigenvector of eigenvalue 1 of the $P^\top$, i.e., that can be obtained in Mathematica via the command

Eigenvectors[(epsilon/5)*[[1,1,1,1,1],[1,1,1,1,1],[1,1,1,1,1],[1,1,1,1,1],[1,1,1,1,1]]
  +(1-epsilon)*0.6,0,0,0,0,0,1,0.2,0,0,0,0,0.8,0,0,0.2,0,0,1,0.2,0.2,0,0,0,0.8].

We can also check that

$$\pi_a(\varepsilon) + \pi_b(\varepsilon) + \pi_c(\varepsilon) + \pi_d(\varepsilon) + \pi_e(\varepsilon)$$

$$= \frac{\varepsilon}{2 + 3\varepsilon} + \frac{2 + 3\varepsilon}{5(1 + 4\varepsilon)} + \frac{\varepsilon}{1 + 4\varepsilon} + \frac{3 + 2\varepsilon}{5(1 + 4\varepsilon)} + \frac{\varepsilon(3 + 2\varepsilon)}{(1 + 4\varepsilon)(2 + 3\varepsilon)}$$

$$= \frac{5\varepsilon(1 + 4\varepsilon)}{5(2 + 3\varepsilon)(1 + 4\varepsilon)} + \frac{(2 + 3\varepsilon)^2}{5(1 + 4\varepsilon)(2 + 3\varepsilon)} + \frac{5\varepsilon(2 + 3\varepsilon)}{5(1 + 4\varepsilon)(2 + 3\varepsilon)}$$

$$+ \frac{(3 + 2\varepsilon)(2 + 3\varepsilon)}{5(2 + 3\varepsilon)(1 + 4\varepsilon)} + \frac{5\varepsilon(3 + 2\varepsilon)}{5(1 + 4\varepsilon)(2 + 3\varepsilon)}$$

$$= \frac{5\varepsilon(1 + 4\varepsilon) + (2 + 3\varepsilon)(5 + 10\varepsilon) + 5\varepsilon(3 + 2\varepsilon)}{5(1 + 4\varepsilon)(2 + 3\varepsilon)}$$

$$= 1.$$

State Ranking

We are now ready to provide a ranking of the states $\{a, b, c, d, e\}$ based on the limiting and stationary distribution $\pi(\varepsilon)$. We note that

$$\pi_a(\varepsilon) < \pi_c(\varepsilon) < \pi_e(\varepsilon) < \pi_b(\varepsilon) < \pi_d(\varepsilon),$$

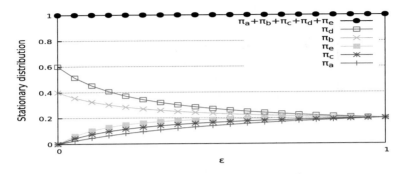

Fig. 8.5 Stationary distribution as a function of $\varepsilon \in [0, 1]$

hence we will rank the states as

Rank	State
1	d
2	b
3	e
4	c
5	a

based on the idea that the most visited states in the long run should rank higher.
In the graph of Fig. 8.5 the stationary distribution is plotted as a function of $\varepsilon \in [0, 1]$.

Convergence Analysis

We note that Assumption (C) p. 169 is satisfied for

$$\theta(\varepsilon) = \frac{\varepsilon}{5\pi_d(\varepsilon)} = \frac{\varepsilon(1 + 4\varepsilon)}{3 + 2\varepsilon},$$

see Fig. 8.6, hence from Proposition 6.21, convergence to the stationary distribution π occurs with speed at least equal to

$$d(n) := \underset{k=1,2,\ldots,N}{\text{Max}} \left\| [P^n]_{k,\cdot} - \pi \right\|_{\text{TV}}$$

$$\leqslant (1 - \theta(\varepsilon))^n$$

$$= \left(\frac{3 + \varepsilon - 4\varepsilon^2}{3 + 2\varepsilon} \right)^n, \qquad n \geqslant 1.$$

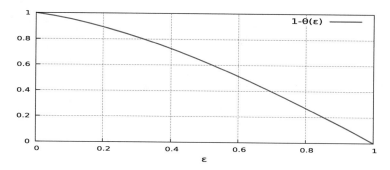

Fig. 8.6 Graph of $1 - \theta(\varepsilon)$ as a function of $\varepsilon \in [0, 1]$

We note that as ε tends to zero, Fig. 8.5 allows us to select a stationary distribution

$$\pi = \lim_{\varepsilon \to 0}[\pi_a(\varepsilon), \pi_b(\varepsilon), \pi_c(\varepsilon), \pi_d(\varepsilon), \pi_e(\varepsilon)] = [0, 0.4, 0, 0.6, 0]$$

which is consistent with Proposition 8.4.

Mean Return Times Analysis

As above, by Theorem 6.6 and Proposition 8.5 we obtain the mean return times for $P(\varepsilon)$, and we note that they are all below $5/\epsilon$. We have

$$\begin{cases} \mu_a(a) = 3 + \dfrac{2}{\varepsilon}, & \mu_b(b) = \dfrac{5(1+4\varepsilon)}{2+3\varepsilon}, & \mu_c(c) = 4 + \dfrac{1}{\varepsilon}, \\[3mm] \mu_d(d) = \dfrac{5(1+4\varepsilon)}{3+2\varepsilon}, & \mu_e(e) = \dfrac{(1+4\varepsilon)(2+3\varepsilon)}{\varepsilon(3+2\varepsilon)}. \end{cases}$$

The remaining of the analysis is similar to that of Sect. 8.4. We have $\lim_{\varepsilon \to 0} \mu_a(a) = \lim_{\varepsilon} \mu_c(c) = \lim_{\varepsilon} \mu_e(e) = +\infty$, and

$$\lim_{\varepsilon \to 0} \mu_b(b) = \frac{5}{2}, \qquad \lim_{\varepsilon \to 0} \mu_d(d) = \frac{5}{3},$$

which do not recover the values $\mu_b(b) = \mu_d(d) = 1$ in case $\varepsilon = 0$. Figure 8.7 plots mean return times as a function of $\varepsilon \in [0, 1]$.

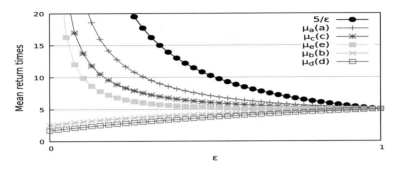

Fig. 8.7 Mean return times as functions of $\varepsilon \in [0, 1]$

Notes

The approach of Bryan and Leise (2006) does not make use of the Markov chain interpretation, and replaces the stationary distribution π with the transposed vector $\pi^\top$ which satisfies the adjoint eigenvalue equation $P^\top \pi^\top = \pi^\top$. See also Liu et al. (2008) for another approach to ranking based on user browsing activity.

Exercises

Problem 8.1 PageRank^{TM} algorithm. We consider the ranking of five web pages a, b, c, d, e which are linked according to the following graph.

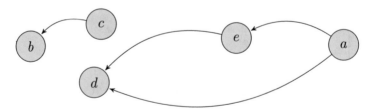

The algorithm works by constructing a self-improving random sequence $(X_n)_{n \geqslant 0}$ which is supposed to "converge" to the best possible search result. Given a search result $X_n = x \in \{a, b, c, d, e\}$, we choose the next search result X_{n+1} with the conditional probability

$$\mathbb{P}(X_{n+1} = y \mid X_n = x) = \frac{1}{n_x}\mathbf{1}_{\{x \to y\}},$$

where n_x denotes the number of outgoing links from x and "$x \to y$" means that x can lead to y in the graph.

(a) Model the process $(X_n)_{n \geq 0}$ as a Markov chain, and find its transition matrix.
(b) Draw the graph of the chain $(X_n)_{n \geq 0}$.
    Is the chain $(X_n)_{n \geq 0}$ reducible?
(c) Does the Markov chain $(X_n)_{n \geq 0}$ admit a limiting distribution independent of the initial state?
(d) Does the Markov chain $(X_n)_{n \geq 0}$ admit a stationary distribution? Find all stationary distribution(s) of the chain $(X_n)_{n \geq 0}$.
(e) In PageRank^{TM} -type algorithms, one typically chooses to perturb the transition matrix P into the new matrix

$$\widetilde{P} := \frac{\varepsilon}{n} \begin{bmatrix} 1 & 1 & 1 & 1 & 1 \\ 1 & 1 & 1 & 1 & 1 \\ 1 & 1 & 1 & 1 & 1 \\ 1 & 1 & 1 & 1 & 1 \\ 1 & 1 & 1 & 1 & 1 \end{bmatrix} + (1 - \varepsilon)P, \qquad \varepsilon \in (0, 1),$$

with $n = 5$ here, where $1 - \varepsilon$ is referred to as the *damping factor*.
Show that $\widetilde{P}$ is a Markov transition matrix and that the corresponding chain $(\widetilde{X}_n)_{n \geq 1}$ is irreducible and aperiodic.
f) Show that $\widetilde{P}$ admits a stationary distribution $\tilde{\pi}$ that satisfies

$$\tilde{\pi} = \left[\frac{\varepsilon}{5}, \frac{\varepsilon}{5}, \frac{\varepsilon}{5}, \frac{\varepsilon}{5}, \frac{\varepsilon}{5} \right] + (1 - \varepsilon)\tilde{\pi} P,$$

and that all probabilities in $\tilde{\pi}$ are greater than $\varepsilon/5$.
g) Compute the stationary distribution of $\widetilde{P}$.
h) Provide a ranking of the states $\{a, b, c, d, e\}$ based on the stationary distribution $\tilde{\pi}$.
(i) Compute the mean return times for $\widetilde{P}$, and show that they are all below $5/\epsilon$.

Chapter 9
Hidden Markov Model

Hidden Markov models attempt to capture hidden sequential information that can be found in data sequences, and belong to the area of unsupervised machine learning. They have numerous applications to clustering, collaborative filtering, recommender systems, computational biology and sequence analysis, genomics, sentiment analysis, natural language processing (NLP), speech and pattern recognition, face recognition, emotion recognition, seismology, climate change studies, finance, etc.

9.1 Graphical Markov Model

In a hidden Markov model, a sequence $(O_k)_{k \in \mathbb{N}}$ of observation is driven by an unknown "hidden" Markov chain $(X_n)_{n \geqslant 0}$ through an emission probability matrix M that encodes the distribution of O_k given the current state of X_k. Our goal is to recover this emission matrix based on the sequence $(O_k)_{k \in \mathbb{N}}$ of observed states.

Hidden Chain

We consider a "hidden" Markov chain $(X_n)_{n \geqslant 0}$ with state space S, transition probability matrix $P = (P_{i,j})_{i,j \in S}$, and initial distribution $\pi = (\pi_i)_{i \in S}$. The Markov chain rule (1.1) can be represented by the graphical Markov model of Fig. 9.1.

© The Author(s), under exclusive license to Springer Nature Switzerland AG 2024
N. Privault, *Discrete Stochastic Processes*, Springer Undergraduate
Mathematics Series, https://doi.org/10.1007/978-3-031-65820-4_9

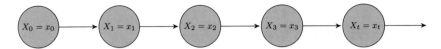

Fig. 9.1 Markovian graphical model

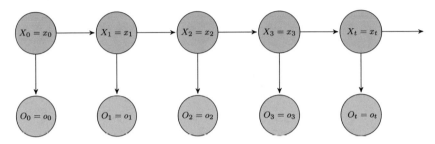

Fig. 9.2 Hidden Markov graphical model

Observed Process

We are observing a process $(O_k)_{k\in\mathbb{N}}$ valued in a set $\mathcal{O}$ of observations. At time $k \in \mathbb{N}$, the state $O_k \in \mathcal{O}$ of the observed process has a conditional distribution given $X_k \in S$ given by the matrix

$$M = (m_{i,j})_{(i,j)\in S\times\mathcal{O}} = (\mathbb{P}(O_t = o \mid X_t = i))_{(i,o)\in S\times\mathcal{O}},$$

called the *emission* probability matrix.

The combined dependency of hidden states and observations can be represented by the graphical Markov model of Fig. 9.2.

The graph of Fig. 9.2 translates into the following dependence relation which will be assumed throughout this chapter:

$$\mathbb{P}(X_t = x_t, \ldots, X_0 = x_0, \ O_t = o_t, \ldots, O_0 = o_0) \tag{9.1}$$

$$= \mathbb{P}(O_t = o_t \mid X_t = x_t) \cdots \mathbb{P}(O_0 = o_0 \mid X_0 = x_0)$$

$$\times \mathbb{P}(X_t = x_t \mid X_{t-1} = x_{t-1}) \cdots \mathbb{P}(X_1 = x_1 \mid X_0 = x_0)\mathbb{P}(X_0 = x_0) \tag{9.2}$$

$$= M_{x_t,o_t} \cdots M_{x_0,o_0} P_{x_{t-1},x_t} \cdots P_{x_0,x_1}\pi_{x_0}, \qquad t \geqslant 0,$$

and together with the chain rule (1.1), it also yields

$$\mathbb{P}(O_t = o_t, \ldots, O_0 = o_0 \mid X_t = x_t, \ldots, X_0 = x_0) \tag{9.3}$$

$$= \mathbb{P}(O_t = o_t \mid X_t = x_t) \cdots \mathbb{P}(O_0 = o_0 \mid X_0 = x_0) = M_{x_t,o_t} \cdots M_{x_0,o_0}, \quad t \geqslant 0.$$

Example

In the case of two hidden states we have $S = \{0, 1\}$ and the hidden two-state chain $(X_n)_{n \geqslant 0}$ has a transition probability matrix of the form:

$$P = \begin{bmatrix} P_{0,0} & P_{0,1} \\ P_{1,0} & P_{1,1} \end{bmatrix} = \begin{bmatrix} \mathbb{P}(X_1 = 0 \mid X_0 = 0) & \mathbb{P}(X_1 = 1 \mid X_0 = 0) \\ \mathbb{P}(X_1 = 0 \mid X_0 = 1) & \mathbb{P}(X_1 = 1 \mid X_0 = 1) \end{bmatrix}$$

with initial distribution

$$\pi = [\pi_0, \pi_1] = [\mathbb{P}(X_0 = 0), \mathbb{P}(X_0 = 1)].$$

In case the set of observations is $\mathcal{O} := \{a, b, c\}$, the conditional distribution of $O_k \in \{a, b, c\}$ given $X_k \in \{0, 1\}$ at every time $k \in \mathbb{N}$ is given by the emission matrix

$$
\begin{aligned}
M &= \begin{bmatrix} M_{0,a} & M_{0,b} & M_{0,c} \\ M_{1,a} & M_{1,b} & M_{1,c} \end{bmatrix} \\
&= \begin{bmatrix} \mathbb{P}(O_k = a \mid X_k = 0) & \mathbb{P}(O_k = b \mid X_k = 0) & \mathbb{P}(O_k = c \mid X_k = 0) \\ \mathbb{P}(O_k = a \mid X_k = 1) & \mathbb{P}(O_k = b \mid X_k = 1) & \mathbb{P}(O_k = c \mid X_k = 1) \end{bmatrix}.
\end{aligned}
$$

This example can be summarized in the graph of Fig. 9.3.

Fig. 9.3 Hidden Markov graph

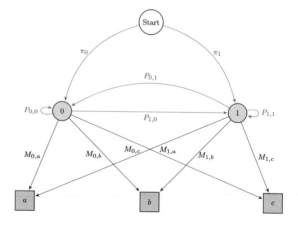

9.2 Forward–Backward Formulas

Proposition 9.1 (Forward Formulas) *For* $t = 1, 2, \ldots, N$ *we have the following identities:*

$$\mathbb{P}(X_t = x_t \mid X_{t-1} = x_{t-1}, \ldots, X_0 = x_0) = \mathbb{P}(X_t = x_t \mid X_{t-1} = x_{t-1}), \qquad (9.4)$$

$$\mathbb{P}(X_t = x_t \mid X_{t-1} = x_{t-1}, \; O_{t-1} = o_{t-1}, \ldots, O_0 = o_0)$$
$$= \mathbb{P}(X_t = x_t \mid X_{t-1} = x_{t-1}) = P_{x_{t-1}, x_t}, \qquad (9.5)$$

$$\mathbb{P}(O_t = o_t \mid X_t = x_t, X_{t-1} = x_{t-1}, O_{t-1} = o_{t-1}, \ldots, O_0 = o_0)$$
$$= \mathbb{P}(O_t = o_t \mid X_t = x_t) = M_{x_t, o_t}. \qquad (9.6)$$

Proof

(i) By summing (9.1) over $o_1, \ldots, o_t \in \mathcal{O}$, we have

$$\mathbb{P}(X_t = x_t, \ldots, X_0 = x_0)$$
$$= \mathbb{P}(X_t = x_t \mid X_{t-1} = x_{t-1}) \cdots \mathbb{P}(X_1 = x_1 \mid X_0 = x_0)\mathbb{P}(X_0 = x_0)$$
$$= \mathbb{P}(X_t = x_t \mid X_{t-1} = x_{t-1})\mathbb{P}(X_{t-1} = x_{t-1}, \ldots, X_0 = x_0),$$

which yields (9.4) and recovers (1.1).

(ii) By (9.1), we have

$$\mathbb{P}(X_t = x_t, \ldots, X_0 = x_0, \; O_t = o_t, \ldots, O_0 = o_0) \qquad (9.7)$$
$$= \mathbb{P}(O_t = o_t \mid X_t = x_t)\mathbb{P}(X_t = x_t \mid X_{t-1} = x_{t-1})$$
$$\times \mathbb{P}(X_{t-1} = x_{t-1}, \ldots, X_0 = x_0, O_{t-1} = o_{t-1}, \ldots, O_0 = o_0),$$

hence by summing over $x_0, x_1, \ldots, x_{t-2} \in S$ and $o_t \in \mathcal{O}$, we have

$$\mathbb{P}(X_t = x_t, X_{t-1} = x_{t-1}, \; O_{t-1} = o_{t-1}, \ldots, O_0 = o_0) \qquad (9.8)$$
$$= \mathbb{P}(X_t = x_t \mid X_{t-1} = x_{t-1})\mathbb{P}(X_{t-1} = x_{t-1}, O_{t-1} = o_{t-1}, \ldots, O_0 = o_0),$$

which implies (9.5).

(iii) By summing (9.7) over $x_0, x_1, \ldots, x_{t-2} \in S$, we have

$$\mathbb{P}(X_t = x_t, X_{t-1} = x_{t-1}, \; O_t = o_t, \ldots, O_0 = o_0)$$
$$= \mathbb{P}(O_t = o_t \mid X_t = x_t)\mathbb{P}(X_t = x_t \mid X_{t-1} = x_{t-1})$$
$$\times \mathbb{P}(X_{t-1} = x_{t-1}, O_{t-1} = o_{t-1}, \ldots, O_0 = o_0),$$

and from (9.8) we obtain

$$\mathbb{P}(X_t = x_t, X_{t-1} = x_{t-1}, O_t = o_t, \ldots, O_0 = o_0)$$
$$= \mathbb{P}(O_t = o_t \mid X_t = x_t)$$
$$\times \mathbb{P}(X_t = x_t, X_{t-1} = x_{t-1}, O_{t-1} = o_{t-1}, \ldots, O_0 = o_0),$$

hence (9.6) holds.

$\square$

Proposition 9.2 (Backward Formulas) *For* $t = 0, 1, \ldots, N - 1$ *we have the following identities:*

$$\mathbb{P}(O_{t+1} = o_{t+1} \mid X_t = x_t, X_{t+1} = x_{t+1}, O_{t+2} = o_{t+2}, \ldots, O_N = o_N)$$
$$= \mathbb{P}(O_{t+1} = o_{t+1} \mid X_{t+1} = x_{t+1}) = M_{x_{t+1}, o_{t+1}}, \tag{9.9}$$

$$\mathbb{P}(X_t = x_t \mid X_{t+1} = x_{t+1}, \ldots, X_N = x_N, O_{t+2} = o_{t+2}, \ldots, O_N = o_N)$$
$$= \mathbb{P}(X_t = x_t \mid X_{t+1} = x_{t+1}). \tag{9.10}$$

Proof We have

$$\mathbb{P}(X_0 = x_0, \ldots, X_N = x_N, O_0 = o_0, \ldots, O_N = o_N)$$
$$= \mathbb{P}(X_0 = x_0)\mathbb{P}(X_1 = x_1 \mid X_0 = x_0) \cdots \mathbb{P}(X_N = x_N \mid X_{N-1} = x_{N-1})$$
$$\times \mathbb{P}(O_0 = o_0 \mid X_1 = x_1) \cdots \mathbb{P}(O_N = o_N \mid X_N = x_N)$$
$$= \mathbb{P}(X_0 = x_0, \ldots, X_t = x_t, O_0 = o_0, \ldots, O_{t-1} = o_{t-1})$$
$$\times \mathbb{P}(X_{t+1} = x_{t+1} \mid X_t = x_t) \cdots \mathbb{P}(X_N = x_N \mid X_{N-1} = x_{N-1})$$
$$\times \mathbb{P}(O_t = o_t \mid X_t = x_t) \cdots \mathbb{P}(O_N = o_N \mid X_N = x_N).$$

(i) By summing over $x_0, \ldots, x_{t-1}, x_{t+2}, \ldots, x_N$ and $o_1, \ldots, o_t$, we have

$$\mathbb{P}(X_t = x_t, X_{t+1} = x_{t+1}, O_{t+1} = o_{t+1}, \ldots, O_N = o_N)$$
$$= \mathbb{P}(O_{t+1} = o_{t+1} \mid X_{t+1} = x_{t+1})$$
$$\times \mathbb{P}(X_t = x_t, X_{t+1} = x_{t+1}, O_{t+2} = o_{t+2}, \ldots, O_N = o_N),$$

hence (9.9) follows.

(ii) We have

$$\mathbb{P}(X_0 = x_0, \ldots, X_N = x_N, \ O_0 = o_0, \ldots, O_N = o_N)$$

$$= \mathbb{P}(X_0 = x_0)\mathbb{P}(X_1 = x_1 \mid X_0 = x_0) \cdots \mathbb{P}(X_N = x_N \mid X_{N-1} = x_{N-1})$$

$$\times \mathbb{P}(O_0 = o_0 \mid X_1 = x_1) \cdots \mathbb{P}(O_N = o_N \mid X_N = x_N)$$

$$= \frac{\mathbb{P}(X_0 = x_0)}{\mathbb{P}(X_{t+1} = x_{t+1})} \mathbb{P}(X_1 = x_1 \mid X_0 = x_0) \cdots \mathbb{P}(X_{t+1} = x_{t+1} \mid X_t = x_t)$$

$$\times \mathbb{P}(O_0 = o_0 \mid X_1 = x_1) \cdots \mathbb{P}(O_{t+1} = o_{t+1} \mid X_t = x_{t+1})$$

$$\times \mathbb{P}(X_{t+1} = x_{t+1}, \ldots, X_N = x_N, \ O_{t+2} = o_{t+2}, \ldots, O_N = o_N)$$

$$= \frac{\mathbb{P}(X_1 = x_1, X_0 = x_0)}{\mathbb{P}(X_{t+1} = x_{t+1})} \mathbb{P}(X_2 = x_2 \mid X_1 = x_1) \cdots \mathbb{P}(X_{t+1} = l x_{t+1} \mid X_t = x_t)$$

$$\times \mathbb{P}(O_0 = o_0 \mid X_1 = x_1) \cdots \mathbb{P}(O_{t+1} = o_{t+1} \mid X_t = x_{t+1})$$

$$\times \mathbb{P}(X_{t+1} = x_{t+1}, \ldots, X_N = x_N, \ O_{t+2} = o_{t+2}, \ldots, O_N = o_N).$$

By summation over $x_0, \ldots, x_{t-1}$ and $o_0, \ldots, o_{t+1}$, we find

$$\mathbb{P}(X_t = x_t, \ldots, X_N = x_N, \ O_{t+2} = o_{t+2}, \ldots, O_N = o_N)$$

$$= \frac{\mathbb{P}(X_t = x_t, X_{t+1} = x_{t+1})}{\mathbb{P}(X_{t+1} = x_{t+1})}$$

$$\times \mathbb{P}(X_{t+1} = x_{t+1}, \ldots, X_N = x_N, \ O_{t+2} = o_{t+2}, \ldots, O_N = o_N),$$

hence (9.10) follows.

$$\square$$

Proposition 9.3 (Forward–Backward Formula) *For $t = 0, 1, \ldots, N-1$ we have the identity*

$$\mathbb{P}(X_t = x_t, \ O_N = o_N, \ldots, O_0 = o_0) \tag{9.11}$$

$$= \mathbb{P}(O_N = o_N, \ldots, O_{t+1} = o_{t+1} \mid X_t = x_t)\mathbb{P}(X_t = x_t, O_t = o_t, \ldots, O_0 = o_0).$$

Proof From (9.2), we have

$$\mathbb{P}(X_N = x_N, \ldots, X_0 = x_0, \ O_N = o_N, \ldots, O_0 = o_0)$$

$$= \mathbb{P}(O_N = o_N \mid X_N = x_N) \cdots \mathbb{P}(O_0 = o_0 \mid X_0 = x_0)$$

$$\times \mathbb{P}(X_N = x_N \mid X_{N-1} = x_{N-1}) \cdots \mathbb{P}(X_1 = x_1 \mid X_0 = x_0)\mathbb{P}(X_0 = x_0)$$

$$= \mathbb{P}(O_N = o_N \mid X_N = x_N) \cdots \mathbb{P}(O_{t+1} = o_{t+1} \mid X_{t+1} = x_{t+1})$$

$$\times \mathbb{P}(X_N = x_N \mid X_{N-1} = x_{N-1}) \cdots \mathbb{P}(X_{t+1} = x_{t+1} \mid X_t = x_t)$$

$$\times \mathbb{P}(X_t = x_t, \ldots, X_0 = x_0, \ O_t = o_t, \ldots, O_0 = o_0)$$

$$= \frac{1}{\mathbb{P}(X_t = x_t)} \mathbb{P}(X_N = x_N, \ldots, X_t = x_t, \ O_N = o_N, \ldots, O_{t+1} = o_{t+1})$$

$$\times \mathbb{P}(X_t = x_t, \ldots, X_0 = x_0, \ O_t = o_t, \ldots, O_0 = o_0), \quad t = 0, 1, \ldots, N-1,$$

and by summation over $x_1, \ldots, x_{t-1}, x_{t+1}, \ldots, x_n$ we obtain

$$\mathbb{P}(X_t = x_t, \ O_N = o_N, \ldots, O_0 = o_0)$$

$$= \mathbb{P}(X_t = x_t, \ O_N = o_N, \ldots, O_{t+1} = o_{t+1})$$

$$\times \frac{1}{\mathbb{P}(X_t = x_t)} \mathbb{P}(X_t = x_t, O_t = o_t, \ldots, O_0 = o_0) \tag{9.12}$$

$$= \mathbb{P}(O_N = o_N, \ldots, O_{t+1} = o_{t+1} \mid X_t = x_t)\mathbb{P}(X_t = x_t, O_t = o_t, \ldots, O_0 = o_0),$$

$t = 0, 1, \ldots, N-1$, which yields (9.11).   □

By (9.3), the conditional probability of observing $(O_0, O_1, O_2) = (c, a, b)$ given that $(X_0, X_1, X_2) = (1, 1, 0)$ splits as

$$\mathbb{P}((O_0, O_1, O_2) = (c, a, b) \mid (X_0, X_1, X_2) = (1, 1, 0))$$

$$= \mathbb{P}(O_0 = c \mid X_0 = 1)\mathbb{P}(O_1 = a \mid X_1 = 1)\mathbb{P}(O_2 = b \mid X_2 = 0)$$

$$= M_{1,c} M_{1,a} M_{0,b},$$

according to the graphical model of p. 230. Using the matrix entries of π, P and M, we can now compute e.g.

$$\mathbb{P}((X_0, X_1, X_2) = (1, 1, 0)) = \mathbb{P}(X_0 = 1, X_1 = 1, X_2 = 0) = \pi_1 P_{1,1} P_{1,0},$$

by Relation (1.1), and the probability

$$\mathbb{P}\big((O_0, O_1, O_2) = (c, a, b) \text{ and } (X_0, X_1, X_2) = (1, 1, 0)\big)$$

of observing the sequence $(O_0, O_1, O_2) = (c, a, b)$ when $(X_0, X_1, X_2) = (1, 1, 0)$, as

$$\mathbb{P}\big((O_0, O_1, O_2) = (c, a, b) \text{ and } (X_0, X_1, X_2) = (1, 1, 0)\big)$$

$$= \mathbb{P}((O_0, O_1, O_2) = (c, a, b) \mid (X_0, X_1, X_2)$$

$$= (1, 1, 0))\mathbb{P}((X_0, X_1, X_2) = (1, 1, 0))$$

$$= \mathbb{P}(O_0 = c \mid X_0 = 1)\mathbb{P}(O_1 = a \mid X_1 = 1)\mathbb{P}(O_2 = b \mid X_2 = 0)$$

$$\times \mathbb{P}((X_0, X_1, X_2) = (1, 1, 0))$$

$$= \pi_1 P_{1,1} P_{1,0} M_{1,c} M_{1,a} M_{0,b}.$$

Using the law of total probability based on all possible values of (X_0, X_1, X_2) we can also compute the probability $\mathbb{P}((O_0, O_1, O_2) = (c, a, b))$ that the observed sequence is (c, a, b), as

$$\mathbb{P}((O_0, O_1, O_2) = (c, a, b)) \tag{9.13}$$

$$= \sum_{x,y,z \in \{0,1\}} \mathbb{P}\big((O_0, O_1, O_2) = (c, a, b) \text{ and } (X_0, X_1, X_2) = (x, y, z)\big)$$

$$= \sum_{x,y,z \in \{0,1\}} \pi_x P_{x,y} P_{y,z} M_{x,c} M_{y,a} M_{z,b}.$$

9.3 Hidden State Estimation

In this section we take $\pi = [\pi_0, \pi_1] := [0.6, 0.4]$, and

$$P := \begin{bmatrix} 0.7 & 0.3 \\ 0.4 & 0.6 \end{bmatrix}, \quad M := \begin{bmatrix} M_{0,a} & M_{0,b} & M_{0,c} \\ M_{1,a} & M_{1,b} & M_{1,c} \end{bmatrix} = \begin{bmatrix} 0.1 & 0.4 & 0.5 \\ 0.7 & 0.2 & 0.1 \end{bmatrix}.$$

See Fig. 9.4. Next, we compute the probabilities

$$\mathbb{P}(X_1 = 1 \mid (O_0, O_1, O_2) = (c, a, b)), \quad \text{and}$$

$$\mathbb{P}(X_1 = 0 \mid (O_0, O_1, O_2) = (c, a, b)).$$

We have

$$\{X_1 = 1\} = \{(X_0, X_1, X_2) = (0, 1, 0)\} \bigcup \{(X_0, X_1, X_2) = (0, 1, 1)\}$$

Fig. 9.4 Hidden Markov graph

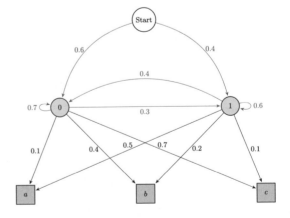

$$\bigcup \{(X_0, X_1, X_2) = (1, 1, 0)\} \bigcup \{(X_0, X_1, X_2) = (1, 1, 1)\}$$

$$= \bigcup_{x,z \in \{0,1\}} \{(X_0, X_1, X_2) = (x, 1, z)\},$$

where the above union is a partition, hence

$$\mathbb{P}(X_1 = 1 \mid (O_0, O_1, O_2) = (c, a, b)) \tag{9.14}$$

$$= \sum_{x,z \in \{0,1\}} \mathbb{P}((X_0, X_1, X_2) = (x, 1, z) \mid (O_0, O_1, O_2) = (c, a, b))$$

$$= \frac{1}{\mathbb{P}((O_0, O_1, O_2) = (c, a, b))}$$

$$\times \sum_{x,z \in \{0,1\}} \mathbb{P}((X_0, X_1, X_2) = (x, 1, z) \text{ and } (O_0, O_1, O_2) = (c, a, b))$$

$$= \frac{1}{\mathbb{P}((O_0, O_1, O_2) = (c, a, b))} \sum_{x,z \in \{0,1\}} \pi_x P_{x,1} P_{1,z} M_{x,c} M_{1,a} M_{z,b}, \tag{9.15}$$

where $\mathbb{P}((O_0, O_1, O_2) = (c, a, b))$ can be computed by (9.13).

Maximum Likelihood Estimation

We can compute the six probabilities

$$\mathbb{P}(X_k = 1 \mid (O_0, O_1, O_2) = (c, a, b)), \quad \mathbb{P}(X_k = 0 \mid (O_0, O_1, O_2) = (c, a, b)),$$

$k = 0, 1, 2$. By proceeding as in (9.14), we find

$$\begin{cases} \mathbb{P}(X_0 = 0 \mid (O_0, O_1, O_2) = (c, a, b)) = 0.825, \\ \mathbb{P}(X_0 = 1 \mid (O_0, O_1, O_2) = (c, a, b)) = 0.175, \\ \mathbb{P}(X_1 = 0 \mid (O_0, O_1, O_2) = (c, a, b)) = 0.256, \\ \mathbb{P}(X_1 = 1 \mid (O_0, O_1, O_2) = (c, a, b)) = 0.744, \\ \mathbb{P}(X_2 = 0 \mid (O_0, O_1, O_2) = (c, a, b)) = 0.636, \\ \mathbb{P}(X_2 = 1 \mid (O_0, O_1, O_2) = (c, a, b)) = 0.364. \end{cases}$$

According to the above estimates, the most likely sequence for (X_0, X_1, X_2) given the observation $(O_0, O_1, O_2) = (c, a, b)$ is

$$(X_0, X_1, X_2) = (0, 1, 0). \tag{9.16}$$

We can also compute the eight probabilities

$$\mathbb{P}\big((X_0, X_1, X_2) = (x, y, z) \text{ and } (O_0, O_1, O_2) = (c, a, b)\big)$$

for all $x, y, z \in \{0, 1\}$, and we identify the most likely sample sequence of values for (X_0, X_1, X_2).

By the results of Sect. 9.1, we find

$$\begin{cases}
\mathbb{P}\big((X_0, X_1, X_2) = (0, 0, 0) \text{ and } (O_0, O_1, O_2) = (c, a, b)\big) = 0.00588, \\
\mathbb{P}\big((X_0, X_1, X_2) = (0, 0, 1) \text{ and } (O_0, O_1, O_2) = (c, a, b)\big) = 0.00126, \\
\mathbb{P}\big((X_0, X_1, X_2) = (0, 1, 0) \text{ and } (O_0, O_1, O_2) = (c, a, b)\big) = 0.0101, \\
\mathbb{P}\big((X_0, X_1, X_2) = (0, 1, 1) \text{ and } (O_0, O_1, O_2) = (c, a, b)\big) = 0.00756, \\
\mathbb{P}\big((X_0, X_1, X_2) = (1, 0, 0) \text{ and } (O_0, O_1, O_2) = (c, a, b)\big) = 0.000448, \\
\mathbb{P}\big((X_0, X_1, X_2) = (1, 0, 1) \text{ and } (O_0, O_1, O_2) = (c, a, b)\big) = 0.0000960, \\
\mathbb{P}\big((X_0, X_1, X_2) = (1, 1, 0) \text{ and } (O_0, O_1, O_2) = (c, a, b)\big) = 0.00269, \\
\mathbb{P}\big((X_0, X_1, X_2) = (1, 1, 1) \text{ and } (O_0, O_1, O_2) = (c, a, b)\big) = 0.00202.
\end{cases}$$

The probability $\mathbb{P}((O_0, O_1, O_2) = (c, a, b))$ that the observed sequence is (c, a, b) is given by (9.13) as

$$\mathbb{P}((O_0, O_1, O_2) = (c, a, b)) = 0.030028 \simeq 3\%. \tag{9.17}$$

From the above computation, we deduce that the sample sequence of values for (X_0, X_1, X_2) which maximizes likelihood given the observation $(O_0, O_1, O_2) = (c, a, b)$ is $(X_0, X_1, X_2) = (0, 1, 0)$, with the probability

$$\mathbb{P}\big((X_0, X_1, X_2) = (0, 1, 0) \text{ and } (O_0, O_1, O_2) = (c, a, b)\big) = 0.0101, \tag{9.18}$$

while the least likely hidden sequence is $(X_0, X_1, X_2) = (1, 0, 1)$, with the probability

$$\mathbb{P}\big((X_0, X_1, X_2) = (1, 0, 1) \text{ and } (O_0, O_1, O_2) = (c, a, b)\big) = 0.0000960.$$

9.4   Forward–Backward Algorithm

Instead of using the formulas

$$\mathbb{P}(O_1, \ldots, O_N) = \sum_{x_1, \ldots, x_N \in S} \pi_{x_1} P_{x_1, x_2} \cdots P_{x_{N-1}, x_N} M_{x_1, O_1} \cdots M_{x_N, O_n}$$

and

$$\mathbb{P}(X_t = x \mid O_1, \ldots, O_N) = \frac{1}{\mathbb{P}((O_0, O_1, O_2) = (c, a, b))}$$

$$\times \sum_{x_1, \ldots, x_{t-1}, x_{t+1}, \ldots, x_N \in S} \pi_{x_1} P_{x_1, x_2} \cdots P_{x_{t-1}, x} P_{x, x_{t+1}} \cdots P_{x_{N-1}, x_N}$$

$$\times M_{x_1, O_1} \cdots M_{x_{t-1}, O_{t-1}} M_{x, O_t} M_{x_{t+1}, O_{t+1}} \cdots M_{x_N, O_N},$$

which have complexity $O(N \times L^N)$ where L is the cardinality of the hidden state space S, we can apply the forward–backward algorithm which instead has complexity $O(L^2 N)$.

Forward Algorithm

Proposition 9.4 *The forward probabilities*

$$\alpha_t(x) := \mathbb{P}(X_t = x, O_1, \ldots, O_t), \qquad t = 1, 2, \ldots, N, \quad x \in S,$$

can be updated by the forward linear recursion

$$\alpha_t(x) = M_{x, O_t} \sum_{y \in S} P_{y, x} \alpha_{t-1}(y), \qquad t = 1, 2, \ldots, N, \quad x \in S,$$

with the initial condition $\alpha_0(x) := \pi_x = \mathbb{P}(X_0 = x)$, $x \in S$.

Proof Using (9.5)–(9.6), for $t \geqslant 1$, we have

$$\alpha_t(x) = \mathbb{P}(X_t = x, O_1, \ldots, O_t)$$

$$= \sum_{y \in S} \mathbb{P}(X_t = x, X_{t-1} = y, O_1, \ldots, O_t)$$

$$= \sum_{y \in S} \mathbb{P}(O_t \mid X_t = x, X_{t-1} = y, O_1, \ldots, O_{t-1})$$

$$\times \mathbb{P}(X_t = x, X_{t-1} = y, O_1, \ldots, O_{t-1})$$

$$= \sum_{y \in S} \mathbb{P}(O_t \mid X_t = x, X_{t-1} = y, O_1, \ldots, O_{t-1})$$

$$\times \mathbb{P}(X_t = x \mid X_{t-1} = y, O_1, \ldots, O_{t-1})$$

$$\times \mathbb{P}(X_{t-1} = y, O_1, \ldots, O_{t-1})$$

$$= \mathbb{P}(O_t \mid X_t = x) \sum_{y \in S} \mathbb{P}(X_t = x \mid X_{t-1} = y) \alpha_{t-1}(y)$$

$$= M_{x,O_t} \sum_{y \in S} P_{y,x} \alpha_{t-1}(y), \qquad t = 1, 2, \ldots, N, \quad x \in S.$$

In addition, we check that with the initial condition $\alpha_0(x) := \pi_x = \mathbb{P}(X_0 = x)$, $x \in S$, we recover

$$\alpha_1(x) = M_{x,O_1} \sum_{y \in S} P_{y,x} \alpha_0(y)$$

$$= M_{x,O_1} \sum_{y \in S} P_{y,x} \mathbb{P}(X_0 = y)$$

$$= \mathbb{P}(O_1 \mid X_1 = x) \mathbb{P}(X_0 = x)$$

$$= \mathbb{P}(X_1 = x, \ O_1).$$

$\square$

Relation (9.17) can be recovered by the forward algorithm using the IPython notebooks

(i) forward_algorithm_tensorflow_page_240.ipynb in Tensorflow, or
(ii) forward_algorithm_pytorch_page_240.ipynb in PyTorch,

uploaded as supplementary material, see also Chapter 6 of Shukla (2018).

Backward Algorithm

Proposition 9.5 *The backward probabilities*

$$\beta_t(x) := \mathbb{P}(O_{t+1}, \ldots, O_N \mid X_t = x), \qquad t = 0, 1, \ldots, N - 1, \quad x \in S.$$

can be updated by the backward linear recursion

$$\beta_t(x) = \sum_{y \in S} M_{y,O_{t+1}} P_{x,y} \beta_{t+1}(y), \qquad t = 0, 1, \ldots, N - 1, \quad x \in S,$$

with the terminal condition $\beta_N(x) := 1, \ x \in S$.

Proof Using (9.9)–(9.10) we have, for $t < N$,

$$\beta_t(x) = \mathbb{P}(O_{t+1}, \ldots, O_N \mid X_t = x)$$

$$= \frac{\mathbb{P}(X_t = x, O_{t+1}, O_{t+2}, \ldots, O_N)}{\mathbb{P}(X_t = x)}$$

$$= \frac{1}{\mathbb{P}(X_t = x)} \sum_{y \in S} \mathbb{P}(X_t = x, X_{t+1} = y, O_{t+1}, \ldots, O_N)$$

$$= \frac{1}{\mathbb{P}(X_t = x)} \sum_{y \in S} \mathbb{P}(O_{t+1} \mid X_t = x, X_{t+1} = y, O_{t+2}, \ldots, O_N)$$

$$\times \mathbb{P}(X_t = x, X_{t+1} = y, O_{t+2}, \ldots, O_N)$$

$$= \frac{1}{\mathbb{P}(X_t = x)} \sum_{y \in S} \mathbb{P}(O_{t+1} \mid X_t = x, X_{t+1} = y, O_{t+2}, \ldots, O_N)$$

$$\times \mathbb{P}(X_t = x \mid X_{t+1} = y, O_{t+2}, \ldots, O_N) \mathbb{P}(X_{t+1} = y, O_{t+2}, \ldots, O_N)$$

$$= \sum_{y \in S} \mathbb{P}(O_{t+1} \mid X_{t+1} = y) \frac{\mathbb{P}(X_t = x \mid X_{t+1} = y)}{\mathbb{P}(X_t = x)} \mathbb{P}(X_{t+1} = y, O_{t+2}, \ldots, O_N)$$

$$= \sum_{y \in S} \mathbb{P}(O_{t+1} \mid X_{t+1} = y) \frac{\mathbb{P}(X_{t+1} = y \mid X_t = x)}{\mathbb{P}(X_{t+1} = y)} \mathbb{P}(X_{t+1} = y, O_{t+2}, \ldots, O_N)$$

$$= \sum_{y \in S} \mathbb{P}(O_{t+1} \mid X_{t+1} = y) \mathbb{P}(X_{t+1} = y \mid X_t = x) \frac{\mathbb{P}(X_{t+1} = y, O_{t+2}, \ldots, O_N)}{\mathbb{P}(X_{t+1} = y)}$$

$$= \sum_{y \in S} \mathbb{P}(O_{t+1} \mid X_{t+1} = y) \mathbb{P}(X_{t+1} = y \mid X_t = x) \mathbb{P}(O_{t+2}, \ldots, O_N \mid X_{t+1} = y)$$

$$= \sum_{y \in S} M_{y, O_{t+1}} P_{x,y} \beta_{t+1}(y), \qquad t = 0, 1, \ldots, N-1, \quad x \in S.$$

In addition, we check that with the terminal condition $\beta_N(x) := 1$, $x \in S$, we recover

$$\beta_{N-1}(x) = \sum_{y \in S} M_{y, O_N} P_{x,y} \beta_N(y)$$

$$= \sum_{y \in S} \mathbb{P}(O_N \mid X_N = y) P_{x,y}$$

$$= \sum_{y \in S} \mathbb{P}(O_N \mid X_N = y, X_{N-1} = x) P_{x,y}$$

$$= \sum_{y \in S} \frac{\mathbb{P}(O_N, \, X_N = y, \, X_{N-1} = x)}{\mathbb{P}(X_N = y, \, X_{N-1} = x)} \mathbb{P}(X_N = y \mid X_{N-1} = x)$$

$$= \sum_{y \in S} \frac{\mathbb{P}(O_N, \, X_N = y, \, X_{N-1} = x)}{\mathbb{P}(X_{N-1} = x)}$$

$$= \frac{\mathbb{P}(O_N, \, X_{N-1} = x)}{\mathbb{P}(X_{N-1} = x)}$$

$$= \mathbb{P}(O_N \mid X_{N-1} = x), \quad x \in S.$$

$\square$

Forward–Backward Algorithm

Proposition 9.6 *For $t = 0, 1, \ldots, N$, we have*

$$\mathbb{P}(X_t = x \mid O_1, \ldots, O_N) = \frac{\alpha_t(x)\beta_t(x)}{\mathbb{P}(O_1, \ldots, O_N)}, \quad x \in S,$$

where

$$\mathbb{P}(O_1, \ldots, O_N) = \sum_{x \in S} \alpha_t(x)\beta_t(x).$$

Proof By (9.11) we have

$$\mathbb{P}(X_t = x, O_1, \ldots, O_N) = \mathbb{P}(O_1, \ldots, O_N \mid X_t = x)\mathbb{P}(X_t = x)$$

$$= \mathbb{P}(O_1, \ldots, O_t \mid X_t = x)\mathbb{P}(O_{t+1}, \ldots, O_N \mid X_t = x)\mathbb{P}(X_t = x)$$

$$= \mathbb{P}(X_t = x, O_1, \ldots, O_t)\mathbb{P}(O_{t+1}, \ldots, O_N \mid X_t = x)$$

$$= \alpha_t(x)\beta_t(x), \qquad t = 1, 2, \ldots, N, \quad x \in S,$$

hence

$$\mathbb{P}(X_t = x \mid O_1, \ldots, O_N) = \frac{\mathbb{P}(X_t = x, O_1, \ldots, O_t)}{\mathbb{P}(O_1, \ldots, O_N)}$$

$$= \frac{\alpha_t(x)\beta_t(x)}{\mathbb{P}(O_1, \ldots, O_N)},$$

where $\mathbb{P}(O_1, \ldots, O_N)$ can be recovered from the normalization condition

$$\sum_{x \in S} \mathbb{P}(X_t = x \mid O_1, \ldots, O_N) = 1,$$

which yields

$$\mathbb{P}(O_1, \ldots, O_N) = \sum_{x \in S} \mathbb{P}(X_t = x, O_1, \ldots, O_N)$$

$$= \sum_{x \in S} \mathbb{P}(O_1, \ldots, O_t, X_t = x) \mathbb{P}(O_{t+1}, \ldots, O_N \mid X_t = x)$$

$$= \sum_{x \in S} \alpha_t(x) \beta_t(x).$$

$\square$

Relation (9.18) can be recovered by the Viterbi algorithm using the IPython notebooks

(i) viterbi_tensorflow_page_243.ipynb in Tensorflow, or
(ii) viterbi_pytorch_page_243.ipynb in PyTorch,

uploaded as supplementary material, see also Chapter 6 of Shukla (2018).

9.5 Baum-Welch Algorithm

Starting from some initial condition $\widehat{\pi}^{(0)}$, $\widehat{P}^{(0)}$, $\widehat{M}^{(0)}$, we build a recursive estimator $\widehat{\pi}^{(n)}$, $\widehat{P}^{(n)}$, $\widehat{M}^{(n)}$ for the model parameters π, P and M, as

$$\widehat{\pi}_i^{(n+1)} := \mathbb{P}^{(n)}(X_0 = i \mid (O_0, O_1, O_2) = (c, a, b)), \tag{9.19a}$$

$$\widehat{P}_{i,j}^{(n+1)} := \frac{\displaystyle\sum_{t=0}^{N-1} \mathbb{P}^{(n)}(X_t = i, X_{t+1} = j \mid (O_0, O_1, O_2) = (c, a, b))}{\displaystyle\sum_{t=0}^{N-1} \mathbb{P}^{(n)}(X_t = i \mid (O_0, O_1, O_2) = (c, a, b))} \tag{9.19b}$$

$$\widehat{M}_{i,k}^{(n+1)} := \frac{\displaystyle\sum_{t=0}^{N} \mathbb{1}_{\{O_t=k\}} \mathbb{P}^{(n)}(X_t = i \mid (O_0, O_1, O_2) = (c, a, b))}{\displaystyle\sum_{t=0}^{N} \mathbb{P}^{(n)}(X_t = i \mid (O_0, O_1, O_2) = (c, a, b))}, \tag{9.19c}$$

where $\mathbb{P}^{(n)}(X_t = i \mid (O_0, O_1, O_2) = (c, a, b))$ is estimated using Proposition 9.6 and $\widehat{\pi}^{(n)}$, $\widehat{P}^{(n)}$, $\widehat{M}^{(n)}$, and similarly for $\mathbb{P}^{(n)}(X_t = i, X_{t+1} = j \mid (O_0, O_1, O_2) = (c, a, b))$. Here, (9.19c) averages the number of times the observed state is "k" given that the hidden state is "i", which gives an estimate of the conditional emission probability $M_{i,k}$.

For example, taking the data of the previous section as initial condition, i.e. $\pi^{(0)} = [\widehat{\pi}_0^{(0)}, \widehat{\pi}_1^{(0)}] := [0.6, 0.4]$, and

$$P^{(0)} := \begin{bmatrix} 0.7 & 0.3 \\ 0.4 & 0.6 \end{bmatrix}, \quad M^{(0)} := \begin{bmatrix} M_{0,a}^{(0)} & M_{0,b}^{(0)} & M_{0,c}^{(0)} \\ M_{1,a}^{(0)} & M_{1,b}^{(0)} & M_{1,c}^{(0)} \end{bmatrix} = \begin{bmatrix} 0.1 & 0.4 & 0.5 \\ 0.7 & 0.2 & 0.1 \end{bmatrix},$$

using (9.19a) we can compute a vector estimate $\widehat{\pi}^{(1)} = [\widehat{\pi}_0^{(1)}, \widehat{\pi}_1^{(1)}]$ as

$$\widehat{\pi}^{(1)} = [\widehat{\pi}_0^{(1)}, \widehat{\pi}_1^{(1)}] = [0.825, 0.175],$$

a matrix estimate using (9.19b) as

$$\widehat{P}^{(1)} = \begin{bmatrix} 0.415 & 0.585 \\ 0.482 & 0.518 \end{bmatrix},$$

and a matrix estimate $\widehat{M}^{(1)}$ using (9.19c) as

$$\widehat{M}^{(1)} = \begin{bmatrix} 0.149 & 0.370 & 0.481 \\ 0.580 & 0.284 & 0.136 \end{bmatrix}.$$

In practice, the Eqs. (9.19a)–(9.19c) are initialized with arbitrary initial values of $\widehat{\pi}$, $\widehat{P}$ and $\widehat{M}$, and then applied iteratively.

Iterating the estimates (9.19a)–(9.19c) is computationally intensive, however this procedure admits an efficient recursive implementation via the Baum-Welch algorithm which is based on the Expectation-Maximization (EM) algorithm, see e.g. Yang et al. (2017) for convergence results for the Baum-Welch algorithm.

Simulation Example

Hidden Markov Model estimation can be implemented by the Baum-Welch algorithm using the IPython notebooks

(i) Baum-Welch_TensorFlow_page_244.ipynb in Tensorflow,
(ii) HMM_PyTorch_page_244.ipynb in PyTorch, or
(iii) HMM_page_244.ipynb,

or the HMM_page_244.R uploaded as supplementary material (Table 9.1).

Table 9.1 Summary of Hidden Markov Models implementations

Package	Tensorflow	PyTorch	hmmlearn (Python)	hmm (R)
Code	Tensorflow*code	PyTorch*code	hmmlearn*code	hmm* code

In the following example we use the HMM (Hidden Markov Model) package in
®️ to estimate the corresponding emission probability matrix M using samples of a
$\{0, 1\}$-valued hidden Markov chain $(X_n)_{n \geqslant 0}$. The source code of the HMM package
is available at https://cran.r-project.org/web/packages/HMM/index.html.

Imagine an alien trying to analyse an English manuscript without any prior
knowledge of English. Using a simple two-state hidden chain $(X_n)_{n \geqslant 0}$ he will try
to uncover some *features* of the language, starting with a *binary classification* of the
English alphabet using the ®️ code_20_page_245.R. A text length of $N \simeq 10,000$
characters can be a minimum. The initial values of π, P and M have to be set
according to random values.

As possible variations, one can try a language different from English, or increase
the state space of $(X_n)_{n \geqslant 0}$ in order to uncover more features of the chosen language.
The estimates of the matrix M obtained from the ®️ code are plotted in Fig. 9.5.

From Fig. 9.5a and b we can infer that the vowels $\{a, e, i, u, o\}$ are more
frequently associated to the state ⓪ of the hidden chain $(X_n)_{n \geqslant 0}$. The vowels
$\{a, e, i, o, u\}$, together with the spacing character "_" amount to 93% of emission
probabilities from state ⓪, and the combined probabilities of vowels from state ①
is only 6.2×10^{-9} %.

Human intervention can be nevertheless required in order to set a probability
threshold that can distinguish vowels from consonants, e.g. to separate "u" from
"t". The classification effect is enhanced in the following Fig. 9.6 that plots $\eta \mapsto$
$(M_{0,\eta}/M_{0,"\_"})((M_{1,"\_"} - M_{1,\eta})/M_{1,"\_"})^2$ by combining the information available in
the two rows of the emission matrix M, showing that "y" recovers its "semi-vowel"
status.

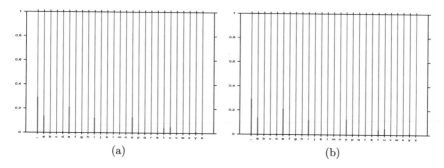

(a)   (b)

Fig. 9.5 Plots of emission probabilities. (**a**) EstimatehmmemissionProbs[1,]. (**b**) EstimatehmmemissionProbs[2,]

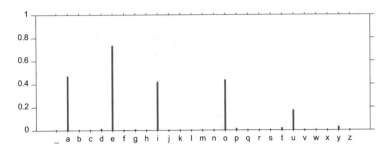

Fig. 9.6 Enhanced classification

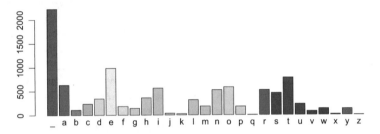

Fig. 9.7 Frequency analysis of alphabet letters

Frequency Analysis

Note that the graphs of Figures 9.5a and 9.5b do *not* represent a frequency analysis.
A frequency analysis of letters can be represented as the histogram of Fig. 9.7 using
the ℝ commands

```
data <- unlist (strsplit (gsub ("[^a-z]", "_", tolower (text)), ""))
barplot(col = rainbow(30), table(data), cex.names=0.7)
```

with the output displayed in Fig. 9.7. The ℝ command

```
estimate$hmm$transProbs
```

yields the estimate of transition probabilities

$$\widehat{P} = \begin{bmatrix} 0 & 1 \\ 0.1906356 & 0.8093644 \end{bmatrix}$$

for the hidden chain $(X_n)_{n \geqslant 0}$. Note that $\widehat{P}$ is not the transition matrix of vowels *vs.* consonants. For example of the word "universities" contains eleven letter transitions $\{un, ni, iv, ve, er, rs, si, it, ti, ie, es\}$, including:

- five vowel-to-consonant transitions $\{un, iv, er, it, es\}$,
- one vowel-to-vowel transition $\{ie\}$,
- four consonant-to-vowel transitions $\{ni, ve, si, ti\}$,
- one consonant-to-consonant transition $\{rs\}$,

which would yield the transition probability estimate

$$\begin{bmatrix} 5/6 & 1/6 \\ 4/5 & 1/5 \end{bmatrix},$$

assuming the alphabet has *already* been partitioned between vowels and consonants. Such a matrix can be estimated on the whole text, from the following ⓡ code:

```
x <- unlist (strsplit (gsub ("[^a-z]", "", tolower (text)), ""))
y <- unlist (strsplit (gsub ("[^a,e,i,o,u]", "2", tolower (x)), ""))
z <- as.numeric(noquote(unlist (strsplit (gsub ("[a,e,i,o,u]", "1",y), ""))))
p <- matrix(nrow = 2, ncol = 2, 0)
for (t in 1:(length(z) - 1)) p[z[t], z[t + 1]] <- p[z[t], z[t + 1]] + 1
for (i in 1:2) p[i, ] <- p[i, ] / sum(p[i, ])
```

This yields

$$\begin{bmatrix} 0.1424749 & 0.8575251 \\ 0.5360502 & 0.4639498 \end{bmatrix},$$

which means that inside the text, a vowel is followed by a consonant for 85.7% of the time, while a consonant is followed by a vowel for 53% of the time.

The Baum-Welch algorithm does more than a simple frequency/transition analysis, as it can estimate the emission probability matrix M, which can be used to partition the alphabet. However, the algorithm is not making a one-to-one association between the states $\{0, 1\}$ of $(X_n)_{n \geqslant 0}$ to letters; the association is only probabilistic and expressed through the estimate $\widehat{M}$ of the emission matrix.

Using a three-state model shows a more definite identification of vowels from state ③ in Fig. 9.8a, and a special weight given to the letters h and t from state ① in Fig. 9.8b.

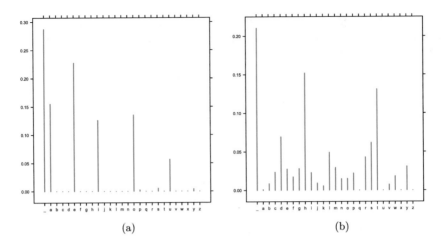

(a)   (b)

Fig. 9.8 Plots of emission probabilities. (**a**) EstimatehmmemissionProbs[3,]. (**b**) EstimatehmmemissionProbs[1,]

Notes

See e.g. Chapter 2 in Stamp (2018), Zucchini et al. (2016) for further reading, Celeux and Durand (2008) for an estimation procedure of the number of hidden states in a hidden Markov model, and Yang et al. (2017) for statistical guarantees for the Baum-Welch algorithm.

Exercises

Exercise 9.1 Consider the graphical hidden Markov model

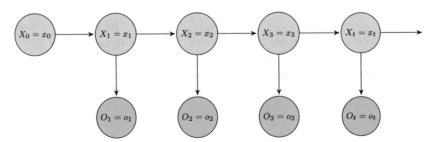

with the relation

$$\mathbb{P}(X_t = i_t, \ldots, X_0 = i_0, O_t = o_t, \ldots, O_1 = o_1)$$
$$= \mathbb{P}(O_t = o_t \mid X_t = i_t) \cdots \mathbb{P}(O_1 = o_1 \mid X_1 = i_1)$$
$$\times \mathbb{P}(X_t = i_t \mid X_{t-1} = i_{t-1}) \cdots \mathbb{P}(X_1 = i_1 \mid X_0 = i_0) \mathbb{P}(X_0 = i_0), \quad t \geq 0.$$

(a) Show that

$$\mathbb{P}(X_t = i_t \mid X_{t-1} = i_{t-1}, \ldots, X_0 = i_0) = \mathbb{P}(X_t = i_t \mid X_{t-1} = i_{t-1}),$$

$t \geq 1$.

(b) Show that

$$\mathbb{P}(X_t = i_t \mid X_{t-1} = i_{t-1}, O_{t-1} = o_{t-1}, \ldots, O_1 = o_1)$$
$$= \mathbb{P}(X_t = i_t \mid X_{t-1} = i_{t-1}), \quad t \geq 1.$$

Exercise 9.2 We consider a two-state hidden Markov chain $(X_n)_{n \geq 0}$ with transition probability matrix $P = (P_{i,j})_{i,j \in S}$ on $S = \{0, 1\}$, in its *stationary distribution* $\pi = (\pi_i)_{i \in S}$. At time $t \geq 0$, the state O_k of an observed process $(O_k)_{k \in \mathbb{N}}$ taking values in a set $\mathcal{O}$ of observations is distributed given $X_k \in \{0, 1\}$ according to the emission matrix $M = (M_{x,o})_{(x,o) \in S \times \mathcal{O}}$, i.e.

$$\mathbb{P}(O_t = o \mid X_t = x) = M_{x,o}, \quad x \in S, \quad o \in \mathcal{O}.$$

(a) Using the identity

$$\mathbb{P}(O_{t+1} = v, O_t = u, X_t = x) = \mathbb{P}(O_{t+1} = v, X_t = x) \mathbb{P}(O_t = u \mid X_t = x),$$

$x = 0, 1$, and the law of total probability, find an expression for the probability

$$\mathbb{P}(O_{t+1} = v, O_t = u), \quad u, v \in \mathcal{O}, \quad t = 0, 1, \ldots, N - 1,$$

using a summation of $\pi_x, P_{x,y}, M_{x,u}, M_{y,v}$ over $x, y \in \{0, 1\}$.

(b) From the result of part (a), find an expression for

$$\mathbb{P}(O_{t+1} \in B, O_t \in A), \quad t = 0, 1, \ldots, N - 1.$$

where A, B are any two subsets of $\mathcal{O}$.

(c) Find expressions for $\mathbb{P}(O_t \in A)$ and

$$\mathbb{P}(O_{t+1} \in B \mid O_t \in A), \quad t = 0, 1, \ldots, N - 1,$$

where A, B are any two subsets of $\mathcal{O}$.

In what follows, we assume that $\mathcal{A}$ and $\mathcal{B}$ form a partition of $\mathcal{O}$, i.e. $\mathcal{O} = \mathcal{A} \cup \mathcal{B}$ and $\mathcal{A} \cap \mathcal{B} = \emptyset$.

(d) Find out and explain how the matrix

$$\begin{bmatrix} \mathbb{P}(O_{t+1} \in \mathcal{A} \mid O_t \in \mathcal{A}) & \mathbb{P}(O_{t+1} \in \mathcal{A} \mid O_t \in \mathcal{B}) \\ \mathbb{P}(O_{t+1} \in \mathcal{B} \mid O_t \in \mathcal{A}) & \mathbb{P}(O_{t+1} \in \mathcal{B} \mid O_t \in \mathcal{B}) \end{bmatrix}$$

compares to

$$P = \begin{bmatrix} P_{0,0} & P_{0,1} \\ P_{1,0} & P_{1,1} \end{bmatrix}$$

whcn

$$\begin{bmatrix} \sum_{u \in \mathcal{A}} M_{0,u} & \sum_{v \in \mathcal{B}} M_{0,v} \\ \sum_{u \in \mathcal{A}} M_{1,u} & \sum_{v \in \mathcal{B}} M_{1,v} \end{bmatrix} = \begin{bmatrix} 1 & 0 \\ 0 & 1 \end{bmatrix}.$$

(e) A numerical experiment classifies $\mathcal{O}$ into a partition $\mathcal{O} = \mathcal{A} \cup \mathcal{B}$ and provides the estimate

$$\widehat{P} = \begin{bmatrix} P_{0,0} & P_{0,1} \\ P_{1,0} & P_{1,1} \end{bmatrix} = \begin{bmatrix} 0.1435747 & 0.8564253 \\ 0.6842348 & 0.3157652 \end{bmatrix}$$

of P. Find the stationary distribution $\pi = [\pi_0, \pi_1]$ of $\widehat{P}$.

(f) The experiment also provides the estimate

$$\begin{bmatrix} \sum_{u \in \mathcal{A}} \widehat{M}_{0,u} & \sum_{v \in \mathcal{B}} \widehat{M}_{0,v} \\ \sum_{u \in \mathcal{A}} \widehat{M}_{1,u} & \sum_{v \in \mathcal{B}} \widehat{M}_{1,v} \end{bmatrix} = \begin{bmatrix} 0.53605372 & 0.4639463 \\ 0.02345197 & 0.9765480 \end{bmatrix}.$$

By applying the result of part (b), find a numerical estimate for the conditional probability matrix

$$\begin{bmatrix} \widehat{\mathbb{P}}(O_{t+1} \in \mathcal{A} \mid O_t \in \mathcal{A}) & \widehat{\mathbb{P}}(O_{t+1} \in \mathcal{A} \mid O_t \in \mathcal{B}) \\ \widehat{\mathbb{P}}(O_{t+1} \in \mathcal{B} \mid O_t \in \mathcal{A}) & \widehat{\mathbb{P}}(O_{t+1} \in \mathcal{B} \mid O_t \in \mathcal{B}) \end{bmatrix}.$$

(g) Compare your numerical answer to part (f) to the actual empirical transition probabilities

$$\begin{bmatrix} 0.1127041 & 0.8872959 \\ 0.2975427 & 0.7024573 \end{bmatrix} \tag{9.20}$$

observed within the data set $\mathcal{O}$ between the subsets $\mathcal{A}$ and $\mathcal{B}$.

Problem 9.3 (Wolfer and Kontorovich (2021)) Consider an *irreducible, reversible,*[1] Markov chain $(X_n)_{n\geq 0}$ admitting a stationary distribution π on the finite state space $S = \{1, 2, \ldots, d\}, d \geq 2$, and started in initial distribution π.

Our goal is to estimate the entries in transition matrix $P = (P_{i,j})_{1\leq i,j\leq d}$ of $(X_n)_{n\geq 0}$ using the estimator

$$\widehat{P}_{i,j}(m) := \frac{1}{N_i(m)} \sum_{k=1}^{m-1} \mathbf{1}_{\{X_k=i, X_{k+1}=j\}}, \qquad i, j = 1, \ldots, d,$$

where

$$N_i(m) := \sum_{k=1}^{m-1} \mathbf{1}_{\{X_k=i\}}$$

denotes the number of returns to state (i) until time $m - 1, i = 1, \ldots, d$.

(a) For any $i = 1, \ldots, d$, we let

$$(Z_i(k))_{k\geq 1} = (Z_i(1), Z_i(2), Z_i(3), \ldots)$$

denote a sequence of independent identically distributed random variables with distribution $P_{i,\cdot}$ on $\{1, \ldots, d\}$, i.e.

$$\mathbb{P}(Z_i(k) = j) = P_{i,j}, \qquad j = 1, \ldots, d, \quad k \geq 1.$$

Show that for all $i = 1, \ldots, d$ we have

$$\mathbb{E}\left[\sum_{j=1}^{d} \left| \frac{1}{n} \sum_{k=1}^{n} \mathbf{1}_{\{Z_i(k)=j\}} - P_{i,j} \right| \right] \leq \sqrt{\frac{d}{n}}, \qquad n \geq 1.$$

Hint. Use Jensen's inequality and the variance of the binomial distribution.

(b) Show that for any $n \geq 1$, the function defined on $\mathbb{R}^n$ by

$$(z(1), \ldots, z(n)) \mapsto \sum_{j=1}^{d} \left| \frac{1}{n} \sum_{k=1}^{n} \mathbf{1}_{\{z(k)=j\}} - P_{i,j} \right|$$

satisfies the bounded differences property with constant $c_i = 2/n, i = 1, \ldots, n$.

---

[1] i.e. $\pi_i P_{i,j} = \pi_j P_{j,i}, i, j = 1, \ldots, d$.

(c) Show that for all $i = 1, \ldots, d$ we have

$$\mathbb{P}\left(\sum_{j=1}^{d}\left|\frac{1}{n}\sum_{k=1}^{n}\mathbf{1}_{\{Z_i(k)=j\}} - P_{i,j}\right| > \varepsilon\right) \leqslant \exp\left(-\frac{n}{2}\operatorname{Max}\left(0, \varepsilon - \sqrt{\frac{d}{n}}\right)^2\right).$$

Hint. Use McDiarmid's inequality.

In what follows, starting from $\widetilde{X}_1$ in the distribution π we let $\widetilde{X}_2 := Z_{\widetilde{X}_1}(1)$, and

$$\widetilde{X}_{k+1} := Z_{\widetilde{X}_k}\left(1 + \widetilde{N}_{\widetilde{X}_k}(k)\right), \qquad k \geqslant 1,$$

where

$$\widetilde{N}_i(k) := \sum_{l=1}^{k-1}\mathbf{1}_{\{\widetilde{X}_l=i\}}, \qquad k \geqslant 1.$$

We also let

$$\widetilde{P}_{i,j}(m) := \frac{1}{\widetilde{N}_i(m)}\sum_{k=1}^{m-1}\mathbf{1}_{\{\widetilde{X}_k=i, \widetilde{X}_{k+1}=j\}}, \qquad i, j = 1, \ldots, d.$$

(d) Show that when $\widetilde{N}_i(m) = n \geqslant 1$ we have

$$\widetilde{P}_{i,j}(m) = \frac{1}{n}\sum_{k=1}^{n}\mathbf{1}_{\{Z_i(k)=j\}}, \qquad i, j = 1, \ldots, d.$$

(e) Show that for $i = 1, \ldots, d$, the distribution of $(\widehat{P}_{i,1}(m), \ldots, \widehat{P}_{i,d}(m))$ on $\{N_i(m) = n\}$ is the same as the distribution of $(\widetilde{P}_{i,1}(m), \ldots, \widetilde{P}_{i,d}(m))$ on $\{\widetilde{N}_i(m) = n\}$.

(f) Show that letting $n_i := \lceil m\pi_i/2 \rceil$, $i = 1, \ldots, d$, for some constant $c_1 > 0$ we have

$$\sum_{n=n_i}^{3n_i}\mathbb{P}\left(\sum_{j=1}^{d}|\widehat{P}_{i,j}(m) - P_{i,j}| > \varepsilon \text{ and } N_i(m) = n\right) \leqslant (2n_i + 1)e^{-c_1 m\pi_i\varepsilon^2},$$

provided that $m \geqslant 4d/(\varepsilon^2\pi_i)$.

(g) Show that

$$\sum_{i=1}^{d}\sum_{n=n_i}^{3n_i}\mathbb{P}\left(\sum_{j=1}^{d}|\widehat{P}_{i,j}(m) - P_{i,j}| > \varepsilon \text{ and } N_i(m) = n\right) \leqslant \frac{2d}{c_1\varepsilon^2}e^{-c_1 m\pi_*\varepsilon^2/2},$$

provided that $m \geqslant 4d/(\varepsilon^2 \pi_*)$ and $\varepsilon \in (0, 1)$, where $\pi_* := \min_{1 \leqslant j \leqslant d} \pi_j$.

Hint. Use the inequality $xe^{-x} \leqslant e^{-x/2}$, $x > 0$.

(h) Show that for all $\varepsilon > 0$ we have

$$
\mathbb{P}\left(\underset{i=1,\ldots,d}{\text{Max}} \sum_{j=1}^{d} \left| \widehat{P}_{i,j}(m) - P_{i,j} \right| > \varepsilon \right)
$$

$$
\leqslant \sum_{i=1}^{d} \sum_{n=n_i}^{3n_i} \mathbb{P}\left(\sum_{j=1}^{d} \left| \widehat{P}_{i,j}(m) - P_{i,j} \right| > \varepsilon \text{ and } N_i(m) = n \right)
$$

$$
+ \mathbb{P}(\exists i \in \{1, \ldots, d\} : N_i(m) \notin [n_i, 3n_i]).
$$

(i) Using the bound in Question (l) of Problem 6.14, show that there exist two constants $c_2, c_3 > 0$ such that

$$
\mathbb{P}(\exists i \in \{1, \ldots, d\} : N_i(m) \notin [n_i, 3n_i]) \leqslant c_2 d e^{-c_3(1-\lambda_1)m\pi_*^2}, \qquad m > 4/\pi_*.
$$

(j) Show that there is a constant $c > 0$ such that for any $\varepsilon, \delta \in (0, 1)$, if

$$
m \geqslant c \, \text{Max}\left(\frac{1}{\varepsilon^2 \pi_*} \, \text{Max}\left(d, \log \frac{d}{\delta\varepsilon} \right), \frac{1}{(1-\lambda_1)\pi_*^2} \log \frac{d}{\delta} \right),
$$

then we have

$$
\mathbb{P}\left(\underset{i=1,\ldots,d}{\text{Max}} \sum_{j=1}^{d} \left| \widehat{P}_{i,j}(m) - P_{i,j} \right| \leqslant \varepsilon \right) \geqslant 1 - \delta.
$$

Chapter 10
Markov Decision Processes

Markov Decision Processes (MDPs) are constructed via the addition of an additional layer of "actions" to a standard Markov model. They are useful to the development of Q-learning algorithms for reinforcement learning. Applications include game theory, recommender systems, robotics, automated control, operations research, information theory, multi-agent systems, swarm intelligence, and genetic algorithms.

10.1 Construction

This section provides the basic construction of Markov decision processes, with some examples.

Definition 10.1 A *Markov Decision Process* (MDP) consists of:

- a state space S,
- a *finite* set $\mathbb{A}$ of possible *actions*,
- a family $(P^{(a)})_{a \in \mathbb{A}}$ of *transition probability matrices* $(P_{i,j}^{(a)})_{i,j \in S}$,
- a state-dependent *reward function* $R : S \to \mathbb{R}$, and
- a state-dependent *policy* $\pi : S \to \mathbb{A}$ which recommends an action $\pi(k) \in \mathbb{A}$ to be taken at any given state in $k \in S$.

When a MDP is in state $X_n = k$ at time n, one looks up the action $a = \pi(k) \in \mathbb{A}$ given by the policy π, and we generate the new value X_{n+1} using the transition probabilities $P_{k,\cdot}^{(\pi(k))} = (P_{k,l}^{(\pi(k))})_{l \in S}$.

In terms of gaming, Markov decision processes represent an evolution from the standard Markov chains that can be used to model board games such as the Snakes and Ladders game. As an example, Markov Decision Processes find a natural application to the Pacman game.

The Tetris game can also be modeled as a Markov decision process.

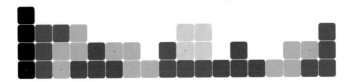

Here, a state consists of a couple

made of one of seven tile shapes and a board configuration. The set of actions consists of the 40 placement choices for the falling tile, and the next state is selected using a new tile shape chosen with uniform probability $1/7$ at each time step.

Example: Deterministic MDP

We consider the deterministic MDP on the state space $S = \{1, 2, 3, 4, 5, 6, 7\}$ with actions $\mathbb{A} = \{\downarrow, \rightarrow\}$ and transition probability matrices

$$
P^{(\downarrow)} := \begin{bmatrix} 0 & 0 & 0 & 1 & 0 & 0 & 0 \\ 0 & 0 & 0 & 0 & 1 & 0 & 0 \\ 0 & 0 & 0 & 0 & 0 & 1 & 0 \\ 0 & 0 & 0 & 1 & 0 & 0 & 0 \\ 0 & 0 & 0 & 0 & 1 & 0 & 0 \\ 0 & 0 & 0 & 0 & 0 & 0 & 1 \\ 0 & 0 & 0 & 0 & 0 & 0 & 1 \end{bmatrix}, \qquad P^{(\rightarrow)} := \begin{bmatrix} 0 & 1 & 0 & 0 & 0 & 0 & 0 \\ 0 & 0 & 1 & 0 & 0 & 0 & 0 \\ 0 & 0 & 1 & 0 & 0 & 0 & 0 \\ 0 & 0 & 0 & 0 & 1 & 0 & 0 \\ 0 & 0 & 0 & 0 & 0 & 1 & 0 \\ 0 & 0 & 0 & 0 & 0 & 0 & 1 \\ 0 & 0 & 0 & 0 & 0 & 0 & 1 \end{bmatrix},
$$

and the reward function $R : S \rightarrow \mathbb{R}$ given by

$$R(1) = 0, \ R(2) = -2, \ R(3) = -1, \ R(4) = -1, \ R(5) = -3, \ R(6) = 5,$$

(10.1)

and $R(7) = 0$.

① $R(1) = 0$	② $R(2) = -2$	③ $R(3) = -1$
④ $R(4) = -1$	⑤ $R(5) = -3$	⑥ $R(6) = +5$

This MDP can be represented by the following graph with state ⑦ as a sink state, where the "⤳" arrows represent the policy choices, while the straight arrows denote Markov transitions.

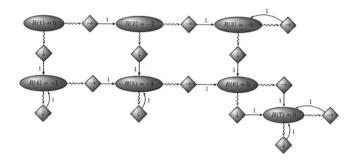

A first look at the above MDP starting from state ① seems to yield

$$\pi(1) = \downarrow, \ \pi(2) = \rightarrow, \ \pi(3) = \downarrow, \ \pi(4) = \rightarrow, \ \pi(5) = \rightarrow$$

as optimal policy, which would ultimately yield a reward $+1$ after starting from state ①. However, a closer look starting from state ⑥ shows that the actual optimal policy is

$$\pi^*(1) = \rightarrow, \ \pi^*(2) = \rightarrow, \ \pi^*(3) = \downarrow, \ \pi^*(4) = \rightarrow, \ \pi^*(5) = \rightarrow,$$

which ultimately yields a reward $+2$ after starting from state ①.

10.2 Reinforcement Learning

The purpose of reinforcement learning is to determine an optimal policy π that maximizes the expected reward function

$$V^\pi(k) := \mathbb{E}_\pi \left[\sum_{n \geq 0} R(X_n) \ \Big| \ X_0 = k \right], \qquad k \in S,$$

where $\mathbb{E}_\pi$ denotes the expectation under a given policy $\pi : S \to \mathbb{A}$. Using first step analysis, we check that the value function $V^\pi(k)$ for a given policy satisfies the equation

$$V^\pi(k) = R(k) + \sum_{l \in S} P_{k,l}^{(\pi(k))} V^\pi(l), \qquad k \in S. \tag{10.2}$$

We also define the action-value functional[1]

$$Q^\pi(k, a) := \mathbb{E}_{\pi,a} \left[\sum_{n \geq 0} R(X_n) \,\Big|\, X_0 = k \right], \qquad k \in S, \quad a \in \mathcal{A}, \tag{10.3}$$

by setting the *first* action at state $\widehat{k}$ to a for a given policy π.

In Proposition 10.2 we show that, similarly to (10.2), the *optimal* action-value function $Q^*(k, a)$, $k \in S$, $a \in \mathbb{A}$, can be written using the transition probability matrix $P^{(a)}$ and the optimal value function $V^*(\cdot)$.[2]

Proposition 10.2 *The action-value functional $Q^\pi(k, a)$ satisfies the equation*

$$Q^\pi(k, a) = R(k) + \sum_{l \in S} P_{k,l}^{(a)} V^\pi(l), \qquad k \in S, \quad a \in \mathbb{A}. \tag{10.4}$$

Proof We have

$$Q^\pi(k, a) := \mathbb{E}_{\pi,a} \left[\sum_{n \geq 0} R(X_n) \,\Big|\, X_0 = k \right]$$

$$= \mathbb{E}_{\pi,a} \left[R(X_0) + \sum_{n \geq 1} R(X_n) \,\Big|\, X_0 = k \right]$$

$$= \mathbb{E}_{\pi,a} \left[R(k) + \sum_{n \geq 1} R(X_n) \,\Big|\, X_0 = k \right]$$

$$= R(k) + \mathbb{E}_{\pi,a} \left[\sum_{n \geq 1} R(X_n) \,\Big|\, X_0 = k \right]$$

---

[1] In the maxima (10.6) the action is taken equal to a at the first step only. After moving to a new state we maximize the future reward according to the best policy choice.

[2] We always assume that $R(\cdot)$ and $(X_n)_{n \geq 0}$ are such that the series in (10.6) converges.

$$= R(k) + \sum_{l \in S} P_{k,l}^{(a)} \mathbb{E}_\pi \left[\sum_{n \geqslant 1} R(X_n) \,\Big|\, X_1 = l \right]$$

$$= R(k) + \sum_{l \in S} P_{k,l}^{(a)} \mathbb{E}_\pi \left[\sum_{n \geqslant 0} R(X_n) \,\Big|\, X_0 = l \right]$$

$$= R(k) + \sum_{l \in S} P_{k,l}^{(a)} V^\pi(l), \qquad k \in S.$$

$\square$

Next, we define the *optimal* value functional $V^*(k)$ as

$$V^*(k) := \underset{\pi}{\mathrm{Max}}\, \mathbb{E}_\pi \left[\sum_{n \geqslant 0} R(X_n) \,\Big|\, X_0 = k \right], \qquad k \in S. \tag{10.5}$$

Similarly, the *optimal* action-value functional $Q^* : S \times \mathcal{A} \longrightarrow \mathbb{R}$ is defined as

$$Q^*(k, a) := \underset{\pi}{\mathrm{Max}}\, \mathbb{E}_{\pi,a} \left[\sum_{n \geqslant 0} R(X_n) \,\Big|\, X_0 = k \right], \qquad k \in S, \quad a \in \mathcal{A}. \tag{10.6}$$

Using first step analysis, we show that the *optimal* action-value functional $Q^*(k, a)$, $k \in S$, $a \in \mathcal{A}$, can be written using the transition probability matrix $P^{(a)}$ and the optimal value functional $V^*(\cdot)$.[3] By an argument similar to that of Proposition 10.3, we have the following result.

Proposition 10.3 *The optimal action-value functional $Q^*(k, a)$ satisfies the inequality*

$$Q^*(k, a) \leqslant R(k) + \sum_{l \in S} P_{k,l}^{(a)} V^*(l), \qquad k \in S, \quad a \in \mathbb{A}.$$

Proof We have

$$Q^*(k, a) := \underset{\pi}{\mathrm{Max}}\, \mathbb{E}_{\pi,a} \left[\sum_{n \geqslant 0} R(X_n) \,\Big|\, X_0 = k \right]$$

---

[3] We always assume that $R(\cdot)$ and $(X_n)_{n \geqslant 0}$ are such that the series in (10.6) is convergent.

$$= \operatorname*{Max}_{\pi} \mathbb{E}_{\pi,a} \left[R(X_0) + \sum_{n \geqslant 1} R(X_n) \,\Big|\, X_0 = k \right]$$

$$= \operatorname*{Max}_{\pi} \mathbb{E}_{\pi,a} \left[R(k) + \sum_{n \geqslant 1} R(X_n) \,\Big|\, X_0 = k \right]$$

$$= R(k) + \operatorname*{Max}_{\pi} \mathbb{E}_{\pi,a} \left[\sum_{n \geqslant 1} R(X_n) \,\Big|\, X_0 = k \right]$$

$$= R(k) + \operatorname*{Max}_{\pi} \sum_{l \in S} P_{k,l}^{(a)} \mathbb{E}_{\pi} \left[\sum_{n \geqslant 1} R(X_n) \,\Big|\, X_1 = l \right]$$

$$\leqslant R(k) + \sum_{l \in S} P_{k,l}^{(a)} \operatorname*{Max}_{\pi} \mathbb{E}_{\pi} \left[\sum_{n \geqslant 1} R(X_n) \,\Big|\, X_1 = l \right]$$

$$= R(k) + \sum_{l \in S} P_{k,l}^{(a)} \operatorname*{Max}_{\pi} \mathbb{E}_{\pi} \left[\sum_{n \geqslant 0} R(X_n) \,\Big|\, X_0 = l \right]$$

$$= R(k) + \sum_{l \in S} P_{k,l}^{(a)} V^*(l), \qquad k \in S.$$

□

In Proposition 10.4, by applying first step analysis we derive the *Bellman equation* satisfied by the *optimal value* function $V^*(k)$.

Proposition 10.4 *The* optimal value *functional V^* satisfies the inequality*

$$V^*(k) \leqslant R(k) + \operatorname*{Max}_{a \in \mathcal{A}} \sum_{l \in S} P_{k,l}^{(a)} V^*(l), \qquad k \in S.$$

Proof For any policy $\pi' : S \to \mathbb{A}$ and $k \in S$, we have

$$\mathbb{E}_{\pi'} \left[\sum_{n \geqslant 0} R(X_n) \,\Big|\, X_0 = k \right] \leqslant \operatorname*{Max}_{a \in \mathcal{A}} \operatorname*{Max}_{\pi} \mathbb{E}_{\pi,a} \left[\sum_{n \geqslant 0} R(X_n) \,\Big|\, X_0 = k \right].$$

Hence, from (10.5) and Proposition 10.3 we obtain

$$V^*(k) = \operatorname*{Max}_{\pi} \mathbb{E}_{\pi} \left[\sum_{n \geqslant 0} R(X_n) \,\Big|\, X_0 = k \right]$$

$$\leqslant \underset{a\in\mathcal{A}}{\text{Max}} \; \underset{\pi}{\text{Max}} \; \mathbb{E}_{\pi,a} \left[\sum_{n\geqslant 0} R(X_n) \; \bigg| \; X_0 = k \right]$$

$$= \underset{a\in\mathcal{A}}{\text{Max}} \; Q^*(k, a)$$

$$= \underset{a\in\mathcal{A}}{\text{Max}} \left(R(k) + \sum_{l\in S} P_{k,l}^{(a)} V^*(l) \right)$$

$$= R(k) + \underset{a\in\mathcal{A}}{\text{Max}} \sum_{l\in S} P_{k,l}^{(a)} V^*(l), \qquad k \in S.$$

$\square$

The equalities

$$V^*(k) = R(k) + \underset{a\in\mathcal{A}}{\text{Max}} \sum_{l\in S} P_{k,l}^{(a)} V^*(l), \qquad k \in S,$$

and

$$Q^*(k, a) = R(k) + \sum_{l\in S} P_{k,l}^{(a)} V^*(l), \qquad k \in S,$$

are called the *Bellman optimal equations*.

Policy Optimization

An optimal policy $\pi^* : S \to \mathbb{A}$ can now be computed from the optimal action-value functional $Q^*(k, a)$, as

$$\pi^*(k) = \text{argmax}_{a\in\mathbb{A}} Q^*(k, a), \quad k \in S. \tag{10.7}$$

Q-Learning

The above optimization problem is solved by a recursive algorithm, starting from an arbitrary initial policy choice $\pi^{(0)}$, and initial data $V^{(0)}(k) = R(k)$, $k \in S$, $a \in \mathcal{A}$. Next, we apply the following steps (i)–(iii) iteratively for $n \geqslant 0$.

(i) *Action-value functional.* Compute $Q^{(n)}(k, a)$ from $V^{(n)}$ using (10.4), for every state $k \in S$ and action $a \in \mathcal{A}$.

(ii) *Policy iteration.* Based on (10.7), apply the policy update

$$\pi^{(n+1)}(k) := \text{argmax}_{a \in \mathbb{A}} Q^{(n)}(k, a), \quad k \in \mathbb{S}.$$

If $\pi^{(n+1)}(k) = \pi^{(n)}(k)$ for all $k \in \mathbb{S}$, then stop.

(iii) *Value iteration.* Update the value function using

$$V^{(n+1)}(k) := \underset{a \in \mathbb{A}}{\text{Max}}\, Q^{(n)}(k, a), \qquad k \in \mathbb{S}.$$

10.3   Example: Deterministic MDP

In the example of Sect. 10.1 we will compute

$$Q^*(k, \downarrow) := \underset{\pi}{\text{Max}}\, \mathbb{E}_{\pi, \downarrow} \left[\sum_{n \geqslant 0} R(X_n) \,\Big|\, X_0 = k \right] \tag{10.8}$$

and

$$Q^*(k, \rightarrow) := \underset{\pi}{\text{Max}}\, \mathbb{E}_{\pi, \rightarrow} \left[\sum_{n \geqslant 0} R(X_n) \,\Big|\, X_0 = k \right], \tag{10.9}$$

starting from state $X_0 = k \in \mathbb{S}$, in the following order: $Q^*(7, \downarrow)$, $Q^*(7, \rightarrow)$, $Q^*(6, \downarrow)$, $Q^*(6, \rightarrow)$, $Q^*(3, \downarrow)$, $Q^*(3, \rightarrow)$, $Q^*(5, \rightarrow)$, $Q^*(5, \downarrow)$, $Q^*(2, \downarrow)$, $Q^*(2, \rightarrow)$, $Q^*(4, \rightarrow)$, $Q^*(4, \downarrow)$, $Q^*(1, \downarrow)$, $Q^*(1, \rightarrow)$.

The optimal action-value functional $Q^*(k, a)$ can be summarized in the graph of Fig. 10.1.

We have

$$Q^*(7, \downarrow) = 0, \quad Q^*(7, \rightarrow) = 0, \quad Q^*(6, \downarrow) = 5, \quad Q^*(6, \rightarrow) = 5,$$

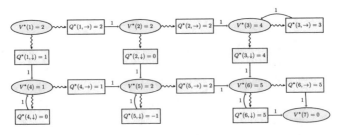

Fig. 10.1 Action-value functional

and $Q^*(3, \downarrow) = 4$. Regarding $Q^*(3, \rightarrow)$, we have

$$Q^*(3, \rightarrow) = -1 + \text{Max}\left(Q^*(3, \downarrow), Q^*(3, \rightarrow)\right),$$

which implies

$$Q^*(3, \rightarrow) < Q^*(3, \downarrow),$$

hence

$$Q^*(3, \rightarrow) = -1 + Q^*(3, \downarrow) = 3.$$

Similarly, we find

$$\begin{cases} Q^*(5, \downarrow) = -1, & Q^*(5, \rightarrow) = 2, \\ Q^*(2, \downarrow) = 0, & Q^*(2, \rightarrow) = 2, \\ Q^*(4, \downarrow) = 0, & Q^*(4, \rightarrow) = 1, \\ Q^*(1, \downarrow) = 1, & Q^*(1, \rightarrow) = 2. \end{cases}$$

We can also solve this system by backward optimization (or dynamic programming), as in the following tree in which optimal policies at each node denoted in green (Fig. 10.2).

Optimal Value Function

Next, we compute the *optimal value* function

$$V^*(k) := \text{Max}_{\pi} \mathbb{E}_{\pi}\left[\sum_{n \geqslant 0} R(X_n) \mid X_0 = k\right],$$

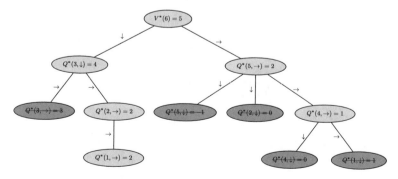

Fig. 10.2 Nodes with optimal and non-optimal policies

at all states $k = 1, 2, \ldots, 7$. At every state $\textcircled{k}$, we have

$$V^*(k) = \text{Max}\left(Q^*(k, \downarrow), Q^*(k, \rightarrow)\right),$$

hence

$$\begin{cases} V^*(7) = 0, \\ V^*(6) = 5, \\ V^*(3) = 4, \\ V^*(5) = 2, \\ V^*(2) = 2, \\ V^*(4) = 1, \\ V^*(1) = 2. \end{cases}$$

The optimal value functional $V^*(k)$, $k = 1, 2, \ldots, 6$, can be summarized in the next table.

$\textcircled{1}\ V^*(1) = 2$	$\textcircled{2}\ V^*(2) = 2$	$\textcircled{3}\ V^*(3) = 4$
$\textcircled{4}\ V^*(4) = 1$	$\textcircled{5}\ V^*(5) = 2$	$\textcircled{6}\ V^*(6) = 5$

The following backward optimization tree is obtained as a subset of the above tree (Fig. 10.3):

Fig. 10.3 Optimal policies

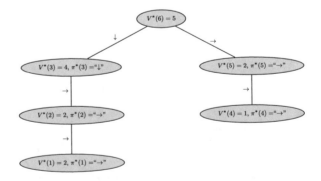

Optimal Policy

We now determine the optimal policy $\pi^* = (\pi^*(1), \pi^*(2), \pi^*(3), \pi^*(4), \pi^*(5))$ of actions leading to the optimal gain starting from any state.[4] We find

$$\pi^* = (\pi^*(1), \pi^*(2), \pi^*(3), \pi^*(4), \pi^*(5), \pi^*(6), \pi^*(7))$$

$$= (\rightarrow, \rightarrow, \downarrow, \rightarrow, \rightarrow, \mathfrak{l}, \mathfrak{l}),$$

which is consistent with output of the ® code_23_page_265.R using the MDPtool-box package.

```
mdp_value_iteration(P,R,discount=1,epsilon=0.01)
$V
[1] 2 2 4 1 2 5 0
$policy
[1] 2 2 1 2 2 1 1
```

The optimal policy $\pi^*(k) \in \{\rightarrow, \downarrow\}$, $k = 1, 2, \ldots, 6$, can be summarized in the next table.

① $\pi^*(1) = $ " $\rightarrow$ "	② $\pi^*(2) = $ " $\rightarrow$ "	③ $\pi^*(3) = $ " $\downarrow$ "
④ $\pi^*(4) = $ " $\rightarrow$ "	⑤ $\pi^*(5) = $ " $\rightarrow$ "	⑥ $\pi^*(6) = $ " $\mathfrak{l}$ "

10.4 Example: Stochastic MDP

Let $p \in [0, 1]$ and consider the stochastic MDP on the state space $\mathbb{S} = \{1, 2, 3, 4, 5, 6, 7\}$, with actions $\mathbb{A} = \{\downarrow, \rightarrow\}$ and transition probability matrices

$$P^{(\downarrow)} := \begin{bmatrix} 0 & 0 & 0 & 1 & 0 & 0 & 0 \\ 0 & 0 & 0 & 0 & 1 & 0 & 0 \\ 0 & 0 & 0 & 0 & 0 & 1 & 0 \\ 0 & 0 & 0 & 1 & 0 & 0 & 0 \\ 0 & 0 & 0 & 0 & 1 & 0 & 0 \\ 0 & 0 & 0 & 0 & 0 & 0 & 1 \\ 0 & 0 & 0 & 0 & 0 & 0 & 1 \end{bmatrix} \qquad P^{(\rightarrow)} := \begin{bmatrix} 0 & 1 & 0 & 0 & 0 & 0 & 0 \\ 0 & 0 & p & 0 & q & 0 & 0 \\ 0 & 0 & 1 & 0 & 0 & 0 & 0 \\ 0 & 0 & 0 & 0 & 1 & 0 & 0 \\ 0 & 0 & 0 & 0 & 0 & 1 & 0 \\ 0 & 0 & 0 & 0 & 0 & 0 & 1 \\ 0 & 0 & 0 & 0 & 0 & 0 & 1 \end{bmatrix},$$

---

[4] The values of $\pi^*(6)$ and $\pi^*(7)$ are not considered because they do not affect the total reward.

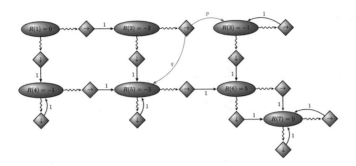

Fig. 10.4 Stochastic MDP

and the reward function (10.1). This MDP can be represented by the following graph with state ⑦ as a sink state, where the ⤳ arrows represent policy choices, while the straight arrows denote Markov transitions (Fig. 10.4).

Using the arguments of Sect. 10.2, we compute the *optimal action-value* function[5]

$$Q^*(k, \downarrow) := \operatorname*{Max}_{\pi} \mathbb{E}_{\pi, \downarrow} \left[\sum_{n \geqslant 0} R(X_n) \mid X_0 = k \right] \tag{10.10}$$

and

$$Q^*(k, \rightarrow) := \operatorname*{Max}_{\pi} \mathbb{E}_{\pi, \rightarrow} \left[\sum_{n \geqslant 0} R(X_n) \mid X_0 = k \right], \tag{10.11}$$

starting from state $X_0 = k \in \mathbb{S}$, in the following order: $Q^*(7, \downarrow)$, $Q^*(7, \rightarrow)$, $Q^*(6, \downarrow)$, $Q^*(6, \rightarrow)$, $Q^*(3, \downarrow)$, $Q^*(3, \rightarrow)$, $Q^*(5, \rightarrow)$, $Q^*(5, \downarrow)$, $Q^*(2, \downarrow)$, $Q^*(2, \rightarrow)$, $Q^*(4, \rightarrow)$, $Q^*(4, \downarrow)$, $Q^*(1, \downarrow)$, $Q^*(1, \rightarrow)$.

Remarks Some values of $Q^*(k, \downarrow)$, $Q^*(k, \rightarrow)$ may now depend on p.

Similarly to the above, we have

$$\begin{cases} Q^*(7, \downarrow) = 0, & Q^*(7, \rightarrow) = 0, \\ Q^*(6, \downarrow) = 5, & Q^*(6, \rightarrow) = 5, \\ Q^*(3, \downarrow) = 4, & Q^*(3, \rightarrow) = 3, \\ Q^*(5, \downarrow) = -1, & Q^*(5, \rightarrow) = 2. \end{cases}$$

---

[5] In the maxima (10.10) the action is taken equal to "$\downarrow$", resp. "$\rightarrow$" at the first step only.

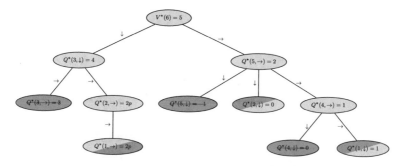

Fig. 10.5 Nodes with optimal and non-optimal policies

We also have $Q^*(2, \downarrow) = 0$ and Proposition 10.3 shows that

$$
\begin{aligned}
Q^*(2, \rightarrow) &= -2 + p \operatorname{Max}\left(Q^*(3, \downarrow), Q^*(3, \rightarrow)\right) + q \operatorname{Max}\left(Q^*(5, \rightarrow), Q^*(5, \rightarrow)\right) \\
&= -2 + p Q^*(3, \downarrow) + q Q^*(5, \rightarrow) \\
&= -2 + 4p + 2q = 2p,
\end{aligned}
$$

and

$$
Q^*(4, \downarrow) = 0, \quad Q^*(4, \rightarrow) = 1, \quad Q^*(1, \downarrow) = 1, \quad Q^*(1, \rightarrow) = Q^*(2, \rightarrow) = 2p.
$$

In other words, we have the following backward optimization (or dynamic programming) tree, in which the choice of colors depends on the position of $p \in (0, 1)$ with respect to the threshold $1/2$ (Fig. 10.5).
Next, using Proposition 10.4 we compute the *optimal value* function

$$
V^*(k) := \operatorname{Max}_{\pi} \mathbb{E}_{\pi}\left[\sum_{n \geqslant 0} R(X_n) \,\Big|\, X_0 = k\right],
$$

at all states $k = 1, 2, \ldots, 7$, depending on the value of $p \in [0, 1]$. At every state $\textcircled{k}$ we have

$$
V^*(k) = \operatorname{Max}\left(Q^*(k, \downarrow), Q^*(k, \rightarrow)\right),
$$

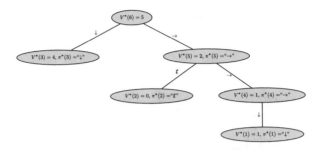

Fig. 10.6 Optimal value function with $p = 0$

hence

$$\begin{cases} V^*(7) = 0, \\ V^*(6) = 5, \\ V^*(5) = 2, \\ V^*(4) = 1, \\ V^*(3) = 4, \\ V^*(2) = 2p, \\ V^*(1) = \text{Max}(2p, 1). \end{cases}$$

Next, we find the optimal policy $\pi^* = (\pi^*(1), \pi^*(2), \pi^*(3), \pi^*(4), \pi^*(5))$ of actions leading to the optimal gain starting from any state, depending on the value of $p \in [0, 1]$.[6]

When $p = 0$, we find

$$\pi^* = (\pi^*(1), \pi^*(2), \pi^*(3), \pi^*(4), \pi^*(5), \pi^*(6), \pi^*(7)) = (\downarrow, \nwarrow, \downarrow, \rightarrow, \rightarrow, \nwarrow, \nwarrow).$$

See Fig. 10.6. The package MDPtoolbox can be used to check our results using the ℝ code_24_page_268.R, as follows.

```
mdp_value_iteration(P,R,discount=1)
$V
[1] 1 0 4 1 2 5 0
$policy
[1] 1 1 1 2 2 1 1
```

When $0 < p < 1/2$, we obtain

$$\pi^* = (\pi^*(1), \pi^*(2), \pi^*(3), \pi^*(4), \pi^*(5), \pi^*(6), \pi^*(7)) = (\downarrow, \rightarrow, \downarrow, \rightarrow, \rightarrow, \nwarrow, \nwarrow)$$

---

[6] The values of $\pi^*(6)$ and $\pi^*(7)$ are not considered here, because they do not affect the total reward.

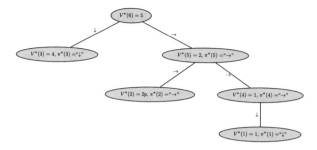

Fig. 10.7 Optimal value function with $0 < p < 1/2$

Fig. 10.8 Optimal value function with $p = 1/2$

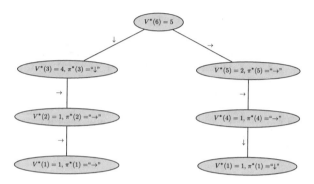

see Fig. 10.7, which is consistent with the output of the ℝ code_24_page_268.R, here with $p = 0.25$:

```
1   $V
    [1] 1.0 0.5 4.0 1.0 2.0 5.0 0.0
3   $policy
    [1] 1 2 1 2 2 1 1
```

When $p = 0.5$, we find

$$\pi^* = (\pi^*(1), \pi^*(2), \pi^*(3), \pi^*(4), \pi^*(5), \pi^*(6), \pi^*(7)) = (\Updownarrow, \rightarrow, \downarrow, \rightarrow, \rightarrow, \Updownarrow, \Updownarrow)$$

see Fig. 10.8, which is consistent with the output of the ℝ code_24_page_268.R, with $p = 0.5$:

```
    $V
2   [1] 1 1 4 1 2 5 0
    $policy
4   [1] 1 2 1 2 2 1 1
```

Fig. 10.9 Optimal value
function with $1/2 < p \leqslant 1$

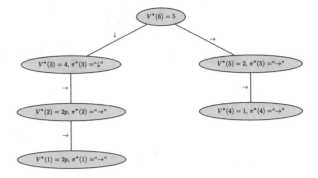

When $1/2 < p \leqslant 1$, we obtain

$$\pi^* = (\pi^*(1), \pi^*(2), \pi^*(3), \pi^*(4), \pi^*(5), \pi^*(6), \pi^*(7))$$

$$= (\rightarrow, \rightarrow, \downarrow, \rightarrow, \rightarrow, \nwarrow, \nwarrow).$$

see Fig. 10.9, which is also consistent with the following output of the ® code_24_
page_268.R, with $p = 0.75$:

```
mdp_value_iteration(P,R,discount=1)
$V
[1] 1.5 1.5 4.0 1.0 2.0 5.0 0.0
$policy
[1] 2 2 1 2 2 1 1
```

Notes

See e.g. Russell and Norvig (1995) for further reading.

Exercises

Exercise 10.1 Consider the Markov chain $(X_n)_{n \geqslant 0}$ on the state space $S = \{a, b, c\}$
whose transition probability matrix P is given by

$$P = \begin{array}{c} \\ a \\ b \\ c \end{array} \begin{array}{ccc} a & b & c \\ \left[\begin{array}{ccc} 1 & 0 & 0 \\ 2/3 & 0 & 1/3 \\ 0 & 1 & 0 \end{array} \right], \end{array}$$

with the following graph:

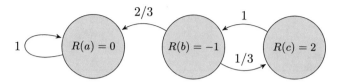

Given the following reward function:

$$R(a) = 0, \qquad R(b) = -1, \qquad R(c) = 2,$$

determine the average accumulated reward $V_a(k) = \mathbb{E}\left[\sum_{n=0}^{\infty} R(X_n) \,\middle|\, X_0 = k\right]$ until the chain is absorbed into state (a) after starting from $k = a, b, c$, assuming a discount factor $\gamma = 1$.

Exercise 10.2 Let $(X_n)_{n \geqslant 0}$ be a three-state Markov chain with the following transition probability graph.

By first step analysis, compute the value function

$$V(k) = \mathbb{E}\left[\sum_{n \geqslant 0} \gamma^n R(X_n) \,\middle|\, X_0 = k\right], \qquad k = 1, 2, 3,$$

where $\gamma \in (0, 1)$ is a discount factor and $R : S \to \mathbb{R}$ is the reward function given by

$$R(1) := -\$2, \quad R(2) := \$3, \quad R(3) := \$1.$$

Exercise 10.3 Let $(X_n)_{n \geqslant 0}$ be a Markov chain with state space S and transition probability matrix $(P_{ij})_{i,j \in S}$. Our goal is to compute the expected value of the infinite discounted series

$$h(i) := \mathbb{E}\left[\sum_{n \geqslant 0} \beta^n c(X_n) \,\middle|\, X_0 = i\right], \qquad i \in S,$$

where $\beta \in (0, 1)$ is the discount coefficient and $c(\cdot)$ is a utility function, starting from state (i).

(a) Show, by a first step analysis argument, that $h(i)$ satisfies the equation

$$h(i) = c(i) + \beta \sum_{j \in S} P_{ij} h(j)$$

for every state $\text{(i)} \in S$.

(b) Consider the Markov chain on the state space $S = \{0, 1, 2\}$ with transition matrix

$$
P = \begin{array}{c c} & \begin{array}{c c c} 0 & 1 & 2 \end{array} \\ \begin{array}{c} 0 \\ 1 \\ 2 \end{array} & \left[\begin{array}{c c c} 0 & 0.5 & 0.5 \\ 0.5 & 0.5 & 0 \\ 0 & 0 & 1 \end{array} \right] \end{array},
$$

and the utility function $c : S \to \mathbb{Z}$ defined by

$$c(0) = \$5, \quad c(1) = -\$2, \quad c(2) = 0.$$

Compute the accumulated utility $h(k)$ after starting from states $k = 0, 1, 2$, by taking $\beta := 1$.

Exercise 10.4 We consider the deterministic Markov Decision Process (MDP) on the state space $S = \{1, 2, \ldots, 10\}$ with actions $\mathbb{A} = \{\downarrow, \to\}$ and reward function $R : S \to \mathbb{R}$ represented in the following graph.

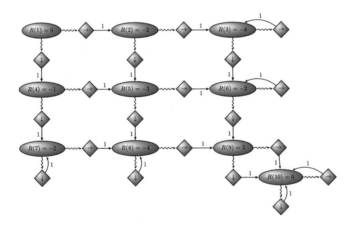

(a) Compute the optimal action-value functional $Q^*(k, a)$, $k = 1, 2, \ldots, 9$, $a \in \{\to, \downarrow\}$.

(b) Compute the optimal value function $V^*(k)$ for $k = 1, 2, \ldots, 9$.

(c) Compute the optimal policy $\pi^*(k) \in \{\to, \downarrow\}$ for $k = 1, 2, \ldots, 9$.

Appendix A
Probability Generating Functions

A.1 Probability Generating Functions

Consider

$$X : \Omega \longrightarrow \mathbb{N} \cup \{+\infty\}$$

a *discrete* random variable possibly taking infinite values. The *probability generating function* (PGF) of X is the *function*

$$G_X : [-1, 1] \longrightarrow \mathbb{R}$$

$$s \longmapsto G_X(s)$$

defined by

$$G_X(s) := \mathbb{E}\big[s^X \mathbb{1}_{\{X<\infty\}}\big] = \sum_{n \geqslant 0} s^n \mathbb{P}(X = n), \qquad -1 \leqslant s \leqslant 1. \tag{A.1}$$

Note that the series summation in (A.1) is over the *finite* integers, which explains the presence of the truncating indicator $\mathbb{1}_{\{X<\infty\}}$ inside the expectation in (A.1).

If the random variable $X : \Omega \longrightarrow \mathbb{N}$ is almost surely finite, i.e. $\mathbb{P}(X < \infty) = 1$, we simply have

$$G_X(s) = \mathbb{E}\big[s^X\big] = \sum_{n \geqslant 0} s^n \mathbb{P}(X = n), \quad -1 \leqslant s \leqslant 1,$$

and for this reason the probability generating function G_X characterizes the *probability distribution* $\mathbb{P}(X = n)$, $n \geqslant 0$, of the random variable $X : \Omega \longrightarrow \mathbb{N}$.

We note that from (A.1) we can write

$$G_X(s) = \mathbb{E}[s^X], \qquad -1 < s < 1,$$

since $s^X = s^X \mathbb{1}_{\{X < \infty\}}$ when $-1 < s < 1$.

Some Properties of Probability Generating Functions

(i) Taking $s = 0$, we have

$$G_X(0) = \mathbb{E}[0^X] = \mathbb{E}[\mathbb{1}_{\{X=0\}}] = \mathbb{P}(X = 0),$$

since $0^0 = 1$ and $0^X = \mathbb{1}_{\{X=0\}}$, hence

$$G_X(0) = \mathbb{P}(X = 0). \tag{A.2}$$

(ii) Taking $s = 1$, we have

$$G_X(1) = \sum_{n \geqslant 0} \mathbb{P}(X = n) = \mathbb{P}(X < \infty) = \mathbb{E}\left[\mathbb{1}_{\{X<\infty\}}\right],$$

hence

$$G_X(1) = \mathbb{P}(X < \infty).$$

(iii) The *derivative* $G'_X(s)$ of $G_X(s)$ with respect to s satisfies

$$G'_X(s) = \sum_{n \geqslant 1} n s^{n-1} \mathbb{P}(X = n), \qquad -1 < s < 1,$$

hence if $\mathbb{P}(X < \infty) = 1$ we have[1]

$$G'_X(1^-) = \mathbb{E}[X] = \sum_{n \geqslant 0} n \mathbb{P}(X = n), \tag{A.3}$$

provided that $\mathbb{E}[X] < \infty$.

---

[1] Here $G'_X(1^-)$ denotes the derivative on the left at the point $s = 1$.

(iv) By computing the second derivative

$$G_X''(s) = \sum_{n \geqslant n} (n-1)ns^{n-2}\mathbb{P}(X=n)$$

$$= \sum_{n \geqslant 0} (n-1)ns^{n-2}\mathbb{P}(X=n)$$

$$= \sum_{n \geqslant 0} n^2 s^{n-2}\mathbb{P}(X=n) - \sum_{n \geqslant 0} ns^{n-2}\mathbb{P}(X=n), \qquad -1 < s < 1,$$

we similarly find

$$G_X''(1^-) = \sum_{n \geqslant 0} (n-1)n\mathbb{P}(X=n)$$

$$= \sum_{n \geqslant 0} n^2 \mathbb{P}(X=n) - \sum_{n \geqslant 0} n\mathbb{P}(X=n)$$

$$= \mathbb{E}[X^2] - \mathbb{E}[X]$$

$$= \mathbb{E}[X(X-1)],$$

provided that $\mathbb{E}[X^2] < \infty$.

More generally, using the *n-th* derivative of G_X we can compute the *factorial moment*

$$G_X^{(n)}(1^-) = \mathbb{E}[X(X-1)\cdots(X-n+1)], \qquad n \geqslant 1, \tag{A.4}$$

provided that $\mathbb{E}[|X^n|] < \infty$. In particular, we have

$$\mathrm{Var}[X] = G_X''(1^-) + G_X'(1^-)(1 - G_X'(1^-)), \tag{A.5}$$

provided that $\mathbb{E}[X^2] < \infty$.

(v) When $X : \Omega \longrightarrow \mathbb{N}$ and $Y : \Omega \longrightarrow \mathbb{N}$ are two finite independent random variables we have

$$G_{X+Y}(s) = \mathbb{E}[s^{X+Y}] \tag{A.6}$$

$$= \mathbb{E}[s^X s^Y]$$

$$= \mathbb{E}[s^X]\mathbb{E}[s^Y]$$

$$= G_X(s)G_Y(s), \qquad -1 \leqslant s \leqslant 1.$$

(vi) The probability generating function can also be used from (A.1) to recover the distribution of the discrete random variable X as

$$\mathbb{P}(X = n) = \frac{1}{n!} \frac{\partial^n}{\partial s^n} G_X(s)_{|s=0}, \qquad n \in \mathbb{N}, \qquad (A.7)$$

extending (A.2) to all $n \geqslant 0$.

Appendix B
Some Useful Identities

Here we present a summary of algebraic identities that are used in this text.

Indicator functions

$$\mathbb{1}_A(x) = \begin{cases} 1 & \text{if } x \in A, \\ 0 & \text{if } x \notin A. \end{cases} \qquad \mathbb{1}_{[a,b]}(x) = \begin{cases} 1 & \text{if } a \leqslant x \leqslant b, \\ 0 & \text{otherwise.} \end{cases}$$

Binomial coefficients

$$\binom{n}{k} := \frac{n!}{(n-k)!k!}, \qquad k = 0, 1, \ldots, n.$$

Exponential series

$$e^x = \sum_{n \geqslant 0} \frac{x^n}{n!}, \qquad x \in \mathbb{R}. \tag{B.1}$$

Geometric sum

$$\sum_{k=0}^{n} r^k = \frac{1 - r^{n+1}}{1 - r}, \qquad r \neq 1. \tag{B.2}$$

Geometric series

$$\sum_{k \geqslant 0} r^k = \frac{1}{1 - r}, \qquad -1 < r < 1. \tag{B.3}$$

Differentiation of geometric series

$$\sum_{k \geqslant 1} k r^{k-1} = \frac{\partial}{\partial r} \sum_{k \geqslant 0} r^k = \frac{\partial}{\partial r} \frac{1}{1 - r} = \frac{1}{(1 - r)^2}, \qquad -1 < r < 1. \tag{B.4}$$

© The Author(s), under exclusive license to Springer Nature Switzerland AG 2024
N. Privault, *Discrete Stochastic Processes*, Springer Undergraduate
Mathematics Series, https://doi.org/10.1007/978-3-031-65820-4

Binomial identity

$$\sum_{k=0}^{n} \binom{n}{k} a^k b^{n-k} = (a + b)^n.$$

Taylor expansion

$$(1 + x)^\alpha = \sum_{k \geqslant 0} \frac{x^k}{k!} \alpha(\alpha - 1) \times \cdots \times (\alpha - (k - 1)). \tag{B.5}$$

References

Agapie, A., & Höns, R. (2007). Analysis of a voter model. *Mathematical Reports (Bucuresti)*, *9*(59)(2), 135–145. (Cited on page 206)

Aldous, D., & Diaconis, P. (1986). Shuffling cards and stopping times. *The American Mathematical Monthly*, *93*(5), 333–348. (Cited on pages 179 and 187)

Althoen, S., King, L., & Schilling, K. (1993). How long is a game of Snakes and Ladders? *The Mathematical Gazette*, *77*(478), 71–76.

Antal, T., & Redner, S. (2005). The excited random walk in one dimension. *Journal of Physics A*, *38*(12), 2555–2577. (Cited on pages 117, 118, 127, 136, and 148)

Asmussen, S. (2003). *Applied probability and queues*. Applications of Mathematics (New York). Stochastic modelling and applied probability. (Vol. 51, 2nd ed.). Springer-Verlag,

Azais, R., & Bouguet, F. (Eds.) (2018). *Statistical inference for piecewise-deterministic Markov processes*. Wiley-ISTE. (Cited on page 41)

Barbu, A., & Zhu, S.-C. (2020). *Monte Carlo methods*. Springer. (Cited on page 206)

Benjamini, I., & Wilson, D. (2003). Excited random walk. *Electronic Communications in Probability*, *8*, 86–92. (Cited on pages 127 and 148)

Besag, J. (1974). Spatial interaction and the statistical analysis of lattice systems. *Journal of the Royal Statistical Society Series B*, *36*, 192–236. (Cited on page 203)

Bhattacharya, B., & Mukherjee, S. (2018). Inference in Ising models. *Bernoulli*, *24*(1), 493–525. (Cited on page 203)

Billingsley, P. (1961). *Statistical inference for Markov processes*. Statistical Research Monographs (Vol. II). University of Chicago Press. (Cited on page 41)

Bosq, D., & Nguyen, H. (1996). *A course in stochastic processes: Stochastic models and statistical inference*. Mathematical and statistical methods. Kluwer. (Cited on page 159)

Bouneffouf, D., Rish, I., & Aggarwal, C. (2020). A survey on practical applications of multi-armed and contextual bandits. In *IEEE congress on evolutionary computation (CEC)* (pp. 1–8). (Cited on page 115)

Broemeling, L. (2018). *Bayesian inference for stochastic processes*. CRC Press. (Cited on page 41)

Bryan, K., & Leise, T. (2006). The $25,000,000,000 eigenvector: The linear algebra behind Google. *SIAM Review*, *48*(3), 569–581. (Cited on pages 173, 180, 213, 215, and 226)

Celeux, G., & Durand, J.-B. (2008). Selecting hidden Markov model state number with cross-validated likelihood. *Computational Statistics*, *23*, 541–564. (Cited on page 248)

Champion, W., Mills, T., & Smith, S. (2007). Lost in space. *Mathematics Scientist*, *32*, 88–96. (Cited on pages 96 and 98)

© The Author(s), under exclusive license to Springer Nature Switzerland AG 2024
N. Privault, *Discrete Stochastic Processes*, Springer Undergraduate
Mathematics Series, https://doi.org/10.1007/978-3-031-65820-4

Chen, M.-F. (2004). *From Markov chains to non-equilibrium particle systems* (2nd ed.). World Scientific Publishing Co. (Cited on page 45)

Chen, B., & Hong, Y. (2012). Testing for the Markov property in time series. *Econometric Theory*, *28*, 130–178. (Cited on page 41)

Chewi, S. (2017). Wald's identity. https://inst.eecs.berkeley.edu/~ee126/fa17/wald.pdf. Accessed Aug 25, 2022. (Cited on page 184)

Diaconis, P. (2009). The Markov chain Monte Carlo revolution. *Bulletin of the American Mathematical Society (N.S.)*, *46*(2), 179–205. (Cited on page 159)

Foucart, S. (2010). Linear algebra and matrix analysis, Lecture 6. https://www.math.drexel.edu/~foucart/TeachingFiles/F12/M504Lect6.pdf. Accessed Aug 25, 2022. (Cited on pages 183, 192, and 193)

Freedman, D. (1983). *Markov chains*. Springer-Verlag. (Cited on page 184)

Goldberg, S. (1986). *Introduction to difference equations* (2nd ed.). Dover Publications, Inc. With illustrative examples from economics, psychology, and sociology. (Cited on page 10)

Gusev, V. (2014). Synchronizing automata with random inputs. In *Developments in language theory* Lecture notes in computer science (Vol. 8633, pp. 68–75). Springer. (Cited on page 79)

Hairer, M. (2016). Convergence of Markov processes. *Lecture Notes, 18*(16), 13. (Cited on page 114)

Jonasson, J. (2009). The mathematics of card shuffling. http://www.math.chalmers.se/~jonasson/convrates.pdf. Accessed Oct 10, 2019. (Cited on pages 179 and 187)

Karlin, S., & Taylor, H. (1981). *A second course in stochastic processes*. Academic Press Inc. (Cited on pages 10 and 14)

Karlin, S., & Taylor, H. (1998). *An introduction to stochastic modeling* (3rd ed.). Academic Press, Inc. (Cited on pages 153 and 158)

Kato, T. (1995). *Perturbation theory for linear operators* (Reprint of the 1980 edition). Classics in Mathematics. Springer-Verlag. (Cited on page 193)

Kijima, M. (1997). *Markov processes for stochastic modeling*. Stochastic Modeling Series. Chapman & Hall. (Cited on page 31)

Latouche, G., & Ramaswami, V. (1999). *Introduction to matrix analytic methods in stochastic modeling*. ASA-SIAM series on statistics and applied probability. Society for Industrial and Applied Mathematics (SIAM)/American Statistical Association. (Cited on page 59)

Levin, D., Peres, Y., & Wilmer, E. (2009). *Markov chains and mixing times*. American Mathematical Society. With a chapter by James G. Propp and David B. Wilson. (Cited on pages 172, 173, 177, and 189)

Lezaud, P. (1998). Chernoff-type bound for finite Markov chains. *Annals of Applied Probability*, *8*(3), 849–867. (Cited on page 191)

Liu, Y., Gao, B., Liu, T.-Y., Zhang, Y., Ma, Z., He, S., Li, H. (2008). BrowseRank: Letting web users vote for page importance. In *SIGIR '08: Proceedings of the 31st annual international ACM SIGIR conference on Research and development in information retrieval* (pp. 451–458). Association for Computing Machinery, Inc. (Cited on page 226)

Markov, A. A. (1909). Recherches sur un cas remarquable d'épreuves dépendantes. *Acta Mathematica*, *33*, 87–104.

Neuts, F. (1981). *Matrix-geometric solutions in stochastic models*. Johns Hopkins series in the mathematical sciences (Vol. 2). Johns Hopkins University Press. An algorithmic approach. (Cited on page 49)

Pólya, G. (1921). Über eine Aufgabe der Wahrscheinlichkeitsrechnung betreffend die Irrfahrt im Strassenetz. *Mathematische Annalen*, *84*, 149–160. (Cited on page 96)

Privault, N. (2018). *Understanding Markov Chains*. Springer Undergraduate Mathematics Series. Springer, second edition. (Cited on pages 25, 32, 38, 40, 41, 94, 117, 153, 158, and 191)

Privault, N. (2022). *Introduction to stochastic finance with market examples* (2nd ed.). Financial mathematics series. Chapman & Hall/CRC. (Cited on pages 117 and 118)

Redner, S. (2001). *A guide to first-passage processes*. Cambridge University Press. (Cited on pages 117 and 118)

Russell, S., & Norvig, P. (1995). *Artificial intelligence* (3rd ed.). Prentice Hall. (Cited on page 270)

Schalekamp, F., & van Zuylen, A. (2009). Rank aggregation: Together we're strong. In *2009 Proceedings of the eleventh workshop on algorithm engineering and experiments (ALENEX09)* (pp. 38–51). SIAM. (Cited on page 218)

Serfozo, R. (2009). *Basics of applied stochastic processes*. Probability and its applications (New York). Springer-Verlag. (Cited on page 184)

Shukla, N. (2018). *Machine learning with TensorFlow*. Manning Publications. (Cited on pages 240 and 243)

Stamp, M. (2018). Introduction to machine learning with applications in information security (1st ed.). Chapman & Hall/CRC. (Cited on page 248)

Vinay, S., & Kok, P. (2019). Statistical analysis of quantum-entangled-network generation. *Physical Review E, 99*, 042313. (Cited on page 59)

Volkov, M. V. (2008). Synchronizing automata and the Černý conjecture. In *Language and automata theory and applications*. Lecture notes in computer science (Vol. 5196, pp. 11–27). Springer. (Cited on page 79)

Wolfer, G., & Kontorovich, A. (2021). Statistical estimation of ergodic Markov chain kernel over discrete state space. *Bernoulli, 27*(1), 532–553. (Cited on page 251)

Yang, F., Balakrishnan, S., & Wainwright, M. (2017). Statistical and computational guarantees for the Baum-Welch algorithm. *Journal of Machine Learning Research, 18*, 1–53. (Cited on pages 244 and 248)

Zucchini, W., MacDonald, I., & Langrock, R. (2016). *Hidden Markov models for time series*. Monographs on statistics and applied probability (2nd ed.). CRC Press. (Cited on page 248)

Index

© The Author(s), under exclusive license to Springer Nature Switzerland AG 2024
N. Privault, *Discrete Stochastic Processes*, Springer Undergraduate
Mathematics Series, https://doi.org/10.1007/978-3-031-65820-4

Author Index

A
Agapie, A., 206
Aggarwal, C., 115, 184
Aldous, D., 179, 187
Althoen, S.C., 27
Antal, T., 117, 118, 127, 136, 148
Asmussen, S., 31
Azais, R., 41

B
Balakrishnan, S., 244, 248
Barbu, A., 206
Benjamini, I., 118, 127, 148
Besag, J., 203
Bhattacharya, B.B., 203
Billingsley, P., 41
Bosq, D., 158, 159
Bouguet, F., 41
Bouneffouf, D., 115, 184
Broemeling, L.D., 41
Bryan, K., 173, 174, 180, 213, 215, 226

C
Celeux, G., 248
Champion, W.L., 96, 98
Chen, B., 41
Chen, M.-F., 45
Chewi, S., 184

D
Diaconis, P., 159, 179, 187
Durand, J.-B., 248

F
Foucart, S., 183, 192, 193
Freedman, D., 184

G
Gao, B., 226
Goldberg, S., 10
Gusev, V.V., 79

H
Hairer, M., 114
He, S., 226
Hong, Y., 41
Höns, R., 206

J
Jonasson, J., 179, 187

K
Karlin, S., 10, 14, 153, 158
Kato, T., 193
Kijima, M., 31
King, L., 27
Kok, P., 59
Kontorovich, A., 251

L
Langrock, R., 248
Latouche, G., 59
Leise, T., 173, 174, 180, 213, 215, 226

Printed in the United States
by Baker & Taylor Publisher Services